ROCKS & MINERALS

An Illustrated

FIELD GUIDE

BY EVELYN MERVINE, PHD

ILLUSTRATED BY VLAD STANKOVIC

Rocks & Minerals: An Illustrated Field Guide

13-Digit ISBN: 978-1-64643-451-0
10-Digit ISBN: 1-64643-451-X

This book may be ordered by mail from the publisher. Please include $5.99 for postage and handling. Please support your local bookseller first!

Books published by Cider Mill Press Book Publishers are available at special discounts for bulk purchases in the United States by corporations, institutions, and other organizations. For more information, please contact the publisher.

Cider Mill Press Book Publishers
"Where good books are ready for press"
501 Nelson Place
Nashville, Tennessee 37214
cidermillpress.com

Cover and interior design by Melissa Gerber
Typography: Adobe Caslon, Caslon 540, DIN 2014, Eveleth Clean Thin, Fontbox Boathouse Filled

Printed in Malaysia

24 25 26 27 OFF 5 4 3 2

First Edition

FOR MY SON CASPIAN,
MY CURIOUS ROCKHOUNDING PARTNER

CONTENTS

INTRODUCTION

You can tell a story with a stone. If you know a little geology, you can look at the rocks and minerals around you and see incredible things. For example, a few months ago I was walking along the beach with my five-year-old son, and we saw some ordinary-looking gray pebbles. My son was not very interested in the pebbles until I showed him that they could float on water. I explained that these floating pebbles are pumice rocks that were formed in a distant volcano and that they floated because escaping volcanic gases had left holes in the rocks. I also told him that the pebbles had traveled hundreds of miles across the ocean before arriving on the beach. During the long journey, the action of the ocean waves turned the originally blocky-shaped rocks into smooth pebbles. Once he knew their story, my son found the gray pebbles fascinating. Now, he looks for pumice stones every time we go to the beach.

The purpose of this book is to teach you a little about rocks and minerals so that you, too, can start to tell stories with stones. Of course, everyone can appreciate beautiful specimens of rocks and minerals. However, I think you will appreciate them even more once you can identify them and you know a little about how they were formed. In addition, after you know a little geology, I think you may start to appreciate some of the simpler-looking varieties of rocks and minerals, too!

In this book, I describe 50 different rocks and minerals and provide information on their properties, where they occur, how they form, and how to identify them. Additionally, I give some information on where you might be able to find them, with a focus on locations in the United States. Last but not least, I provide a fun fact about each rock and mineral.

There are hundreds of different rocks and thousands of different minerals. Therefore, this book is not intended to be a comprehensive guide. Rather, this book is intended to be a fun tour through the world of geology, introducing you to some of the most common rocks and minerals that you may encounter. In addition, I provide information on a few rare rocks and minerals. You are unlikely to encounter these rocks and minerals, except perhaps in a museum or jewelry shop. However, rare rocks and minerals are incredibly fascinating, and you will enjoy learning about them.

To help you use this book, I provide some basic information about rock and mineral classification and identification below. I also list some tools that geologists use to help them identify rocks and minerals. Additionally, there is a glossary of geologic terms at the back of this book. If you would like to continue your study of geology, I also provide a list of suggested rock and mineral reference books.

ROCK CLASSIFICATION AND IDENTIFICATION

A rock is a naturally occurring solid mass that is comprised of one or more minerals or mineraloids. The Earth's crust (or outer layer), as well as much of its interior, is made out of rocks. Rocks are classified into three types: igneous, sedimentary, and metamorphic. Rocks can be transformed from one type of rock into another type of rock through the rock cycle.

Igneous rocks form from the melting of other rocks. Igneous rocks can form either from magma, which is molten material located underground, or from lava, which is molten material that flows on Earth's surface. Lava is either erupted by volcanoes or extruded through fractures in Earth's crust. Igneous rocks that form from the cooling of magma are called intrusive igneous rocks. Igneous rocks that form from the cooling of lava are called extrusive igneous rocks. They can also be referred to as volcanic rocks.

Sedimentary rocks are composed of fragments of rocks, minerals, and/or organic materials that become cemented together (non-chemical sedimentary rocks) or from minerals that precipitate from solution (chemical sedimentary rocks). Sedimentary rocks are commonly described as being clastic or non-clastic. Clasts are fragments of rocks or minerals that are formed by the weathering of other rocks. These clasts can accumulate as sediments, which can be solidified into sedimentary rock. Clastic sedimentary rocks are often categorized by the size, shape, and composition of their clasts. Non-clastic rocks do not have visible clasts. They are either chemical sedimentary rocks or rocks formed from small organic particles. Sedimentary rocks that form from organic materials are referred to as biological sedimentary rocks.

Metamorphic rocks are formed when rocks are subjected to heat and pressure that transforms them into another type of rock, a process known as metamorphism. During metamorphism, the texture and minerals of the rock change. Importantly, the rock is not completely melted. When rocks are melted, they transform into igneous rocks, not metamorphic rocks.

To identify rocks, you need to look at their properties, such as their colors, mineral compositions, and textures. Color is helpful for identifying many rocks. However, it is important to note that different rocks can have the same color. In addition, it is common for the same rock to come in multiple colors. Looking at the type and proportion of minerals found in a rock is a very good way to identify it. Mineral identification is easiest when the mineral crystals are large enough to see with the naked eye. For some rocks, the mineral crystals are small and you may need to use a hand lens or even a microscope to identify them. Looking at texture is an important way to identify many rocks. The rock descriptions in this book include many terms that geologists use to classify rock textures. Some of these terms, such as "glassy," will be familiar to you. Other terms, such as "phaneritic" and "porphyritic," may seem mystifying at first. However, the terms are explained in the descriptions, and they are also included in the glossary at the back of the book. For some rocks, other properties, such as density and magnetism, may be useful for identification. For rocks that contain carbonate minerals, it can be useful to put a small amount of dilute acid on the rock, since carbonate minerals will fizz when exposed to acid.

MINERAL CLASSIFICATION AND IDENTIFICATION

A mineral is a solid, naturally occurring, inorganic material that has a definite chemical composition (for some minerals, the composition varies within a range) and crystal structure. A mineraloid is similar to a mineral but does not have a crystal structure. An example of a mineraloid is glass, which has an amorphous structure, which means that the atoms are not arranged in a regular crystal pattern.

Minerals are classified by their crystal structures, which is the geometric arrangement of the crystal as defined by its three axes A, B, and C. For example, crystals that form in the cubic system have a cubic structure with the A, B, and C axes having the same length and intersecting at 90-degree angles. The seven crystal systems are: cubic (also called isometric), tetragonal, hexagonal, trigonal, orthorhombic, monoclinic, and triclinic. If you want to learn more about crystal systems, take a look in one of the mineralogy books recommended at the back of this book.

Geologists use a number of properties to identify minerals. Some of the most useful properties for mineral identification are:

COLOR – It can be useful to look at color, and some minerals come in a characteristic color. However, use color with caution because many minerals come in several different colors.

TRANSPARENCY – This is the degree to which light can shine through a mineral. Minerals can be transparent (all light goes through), translucent (some light goes through), or opaque (no light goes through). The transparency of some minerals can vary, so it is not always a useful diagnostic tool.

STREAK – This is the color of fine mineral powder that is left on a streak plate (an unglazed ceramic plate) when a mineral is scraped across the surface of the plate. Streak is a very useful property because a mineral often has the same color streak, even when a mineral comes in multiple colors. For example, calcite can come in several different colors, but it always has a white streak. Note that a streak plate has a hardness of about 7 on the Mohs Hardness Scale, so it can only be used to test the streak of minerals with a hardness of less than 7. Streak can still be a useful way to identify harder minerals, but geologists need to generate mineral powder using other methods, usually in a laboratory.

LUSTER – This is the way in which light interacts with the surface of a mineral. Geologists use a number of terms to describe luster, such as vitreous (glass-like), metallic, silky, pearly, adamantine (diamond-like), and dull.

HARDNESS – Hardness is a very useful property for mineral identification. Geologists commonly use a relative scale of hardness called the Mohs Hardness Scale. This scale is defined by ten characteristic minerals: talc (hardness = 1), gypsum (hardness = 2), calcite (hardness = 3), fluorite (hardness = 4), apatite (hardness = 5), feldspar (hardness = 6), quartz (hardness = 7), topaz (hardness = 8), corundum (hardness = 9), and diamond (hardness = 10). The relative hardness of an unknown mineral can be determined by trying to scratch it with one of the minerals on the Mohs Hardness Scale. If the unknown mineral can be scratched by the known mineral, then it is softer than the reference mineral. If it cannot be scratched, then it is harder than the reference mineral. You can also test mineral hardness using common objects. For example, your fingernail has a hardness of about 2.5 on the Mohs Hardness Scale.

SPECIFIC GRAVITY – This is a dimensionless measurement related to the density of a mineral. Specifically, it is the ratio of the mass of a mineral to the mass of an equal volume of water. For example, the mineral quartz has a specific gravity of about 2.6, which means that 1 cm^3 of quartz will be 2.6 times denser than 1 cm^3 of water. Informally, you can think of specific gravity as an indication of the heaviness of a mineral. The higher the specific gravity, the heavier the mineral.

Geologists also sometimes use additional properties to identify minerals. For example, crystal habit, which is the shape of mineral crystals, can be very useful for identifying many minerals. However, it is important to note that it is common for minerals to form crystals in more than one type of shape. Additionally, minerals only form in characteristic shapes under certain conditions—for example, when they are able to grow into open cavities. Geologists can also observe how a mineral breaks to look for characteristic cleavage planes (planes along which a mineral tends to break) or fracture surfaces. For some minerals, properties such as magnetism, fluorescence, and phosphorescence can be useful for identification. However, starting with the properties listed above will enable you to identify most common minerals.

USEFUL IDENTIFICATION TOOLS

Geologists use a number of tools to help them identify rocks and minerals. When working in the field, geologists often wear a special vest and a special belt to carry these tools. You can also carry these tools in a backpack. You can order these tools online or find them in specialist supply shops.

Some useful tools are:

ROCK HAMMER – This can be used to break open a rock, exposing a fresh surface that is easier to identify. Be sure to wear appropriate safety gear, including safety glasses, when using a rock hammer.

HAND LENS – This is very useful for looking at small mineral crystals. It is a good idea to attach your hand lens to a string so that you can wear it around your neck. You will use this tool often!

MAGNET - This can be used to test if a rock or mineral is magnetic. Geologists often use small pencil-shaped magnets that are easy to carry in the field.

STREAK PLATE - This is a ceramic plate that is used to check the color of a mineral streak.

MOHS HARDNESS TESTING KIT - This is a kit that contains small samples of all of the minerals on the Mohs Hardness Scale (except for diamond!) that can be used to determine the relative hardness of unknown minerals. As an alternative, you can use a few different common items to evaluate hardness, such as your fingernail (hardness of approximately 2.5), a copper penny (hardness of approximately 3.5), a knife (hardness of approximately 5.5), and a steel file (hardness of approximately 6.5).

ACID BOTTLE - This is used to check for carbonate minerals, which will fizz when exposed to acid. Professional geologists usually fill their acid bottles with dilute hydrochloric acid. However, this acid is hazardous and needs to be used with appropriate safety protocols, including wearing gloves. A safer alternative is to fill the bottle with vinegar, which is also a weak acid. However, it is important to note that carbonates will fizz less vigorously with vinegar compared to hydrochloric acid.

FIELD NOTEBOOK - This is useful for recording notes and sketches. Geologists often use a special type of field notebook that has waterproof paper, so that they can take notes in any type of weather, including rain, sleet, and snow!

SAMPLE COLLECTION BAGS - These are special bags for collecting rock and mineral specimens. Geologists often use small cotton bags with drawstrings. Before you collect specimens, be sure that you are permitted to collect them. For example, you are not allowed to collect rocks and minerals in most national parks and public lands. Be sure to carry a permanent marker for writing down the sample name and collection location on the specimen bag.

COMMON ROCKS

GRANITE

ROCK TYPE: Igneous

COLOR: Variable—commonly white, pink, red, black, and gray

MINERALOGY: Primarily feldspar and quartz

TEXTURE: Phaneritic, sometimes prophyritic

DESCRIPTION AND OCCURRENCE: Granite is a phaneritic (coarse-grained), intrusive (formed underground from magma), igneous rock. The mineral crystals (or grains) in granite are large enough to see clearly with the naked eye. Granite consists primarily of the minerals feldspar and quartz. Granite contains two types of feldspar: plagioclase feldspar (generally white in color) and alkali feldspar (variable in color, but often white or pink). In addition to quartz and the two types of feldspar, granite will contain a darker-colored mineral that is most commonly a type of mica (silver muscovite or black biotite) and/or an amphibole. The color of granite varies depending on the colors of the minerals present and the relative amounts of those minerals. Commonly, granite consists of white feldspar with some clear or gray quartz and small amounts of the darker minerals. Granite can look pink or red when the alkali feldspar has this color. Granites often display an equigranular texture, where the grains are all about the same size. However, some granites may have feldspar grains, most commonly alkali feldspar grains, that are much larger than the other mineral grains. This type of granite is called a "porphyritic granite."

Granite is a felsic rock, which means that it has a high silica content. Granite also generally contains a significant amount of aluminum, potassium, and sodium. Some granites contain uranium and thorium, which makes them slightly radioactive! Don't worry, however. Most granites are not radioactive enough to be harmful to humans. However, large volumes of granite bedrock may generate radon gas, which is produced by the radioactive decay reactions. Radon can sometimes accumulate in basements and in groundwater pumped from wells, so radon monitors sometimes need to be installed in these locations.

Granite belongs to a family of similar-looking rocks known as granitoids. For example, granodiorite looks similar to granite but contains a higher amount of plagioclase feldspar relative to alkali feldspar. As another example, syenite also looks similar to granite but has a higher percentage of alkali feldspar relative to plagioclase feldspar and a lower amount of quartz. At a quick glance, it can be challenging to tell different granitoids apart. Geologists identify different granitoids by carefully determining the percentages of the different minerals, often using a hand lens or microscope to assist in mineral identification.

Granite forms under the Earth's surface as intrusions underneath other rocks. Over time, the rocks that were overlying the granite can erode away or the granite can be uplifted by the movement of

Earth's tectonic plates. Granite is a hard, durable rock, making it difficult to erode, which allows it to remain and form mountains. This makes granite a common rock on the surfaces of continents. Granite tends to round when it weathers, so granitic hills or mountains will often have a rounded shape. Similarly, granite boulders tend to be rounded.

In the United States, granite can be found in many locations. The best places to find granite are mountainous regions, such as the Appalachian Mountains, the Rocky Mountains, and the Sierra Nevada Mountains. One great place to see granite is Yosemite National Park in California. The large cliffs in the park, such as El Capitan and Half Dome, are comprised of granite or granitoids. Another great place to see granite is Deer Isle in Maine. The island is composed mostly of granite, which has been quarried since the 1860s and has been used in numerous famous monuments and structures, such as the Statue of Liberty, the John F. Kennedy Gravesite at Arlington National Cemetery, and the Harvard Bridge joining Boston and Cambridge in Massachusetts.

Granite is commonly used as a building stone and a decorative stone. For example, kitchen countertops or decorative building stone foundations are commonly made of granite. It is important to note that the decorative stone industry tends to call any granite-like rock "granite." In reality, a decorative stone labelled "granite" might be a different granitoid or might not even be a granitoid at all. If you want to know if your kitchen countertop is really granite, ask a geologist to take a look at it. However, don't worry if your countertop is not really granite. Many types of rocks make great countertops!

IDENTIFICATION: Granite will have large mineral crystals that are visible with the naked eye. The best way to identify granite is to look for the minerals feldspar and quartz. If you see a significant amount of these minerals, then the rock is most likely a granite or at least a granitoid. Using a hand lens may be helpful to see the minerals more clearly.

FUN FACT: New Hampshire is known as "The Granite State" because there are many granite outcrops in the state. "The Old Man of the Mountain" was a famous natural stone face on a granite cliff near Franconia, New Hampshire, and is often used as a symbol for the state. Unfortunately, "The Old Man of the Mountain" collapsed in 2003. However, the stone face image still appears on New Hampshire license plates to this day.

RHYOLITE

ROCK TYPE: Igneous

COLOR: Variable—often white, pink, light gray, or light brown

MINERALOGY: Mostly quartz and feldspar

TEXTURE: Usually aphanitic, sometimes porphyritic

DESCRIPTION AND OCCURRENCE: Rhyolite is an aphanitic (fine-grained), extrusive (formed from lava that flowed on Earth's surface), igneous rock. Rhyolite is the extrusive equivalent of granite, which means that it has the same chemistry and mineral composition as granite. However, because rhyolite forms by rapid cooling of aboveground lava while granite forms from slow cooling of underground magma, the crystals in rhyolite are much smaller than those found in granite. The mineral crystals (or grains) in rhyolite are often too small to be seen with the naked eye. Like granite, rhyolite consists mostly of quartz and feldspar (both alkali feldspar and plagioclase feldspar) and can also contain mica and amphibole. Sometimes, rhyolite will have a few larger crystals (called "phenocrysts"), which are usually the same minerals that are found as small crystals. When rhyolite contains no or very few phenocrysts, its texture is aphanitic. When rhyolite contains a significant number of phenocrysts, its texture is porphyritic.

The color of rhyolite is variable, but it is commonly white, pink, light gray, or light brown. Rhyolite can be uniform in color. It can also have alternating layers of varying lighter and darker colors. Sometimes, these layers are flow banding, which forms as the lava flows across the Earth's surface.

Rhyolite forms from silica-rich lavas that can erupt explosively. These lavas are generally formed at tectonic plate boundaries and intracontinental rifts (places where continents are splitting apart). Rhyolites commonly occur at volcanoes along convergent plate boundaries where an oceanic plate subducts underneath a continental plate. In the United States, good places to see rhyolite are Yellowstone National Park in Wyoming and Montana, Mammoth Mountain and Long Valley Caldera in California, and Valles Caldera National Preserve in New Mexico.

IDENTIFICATION: If you see a light-colored, fine-grained rock that is either uniform in color or has banding, it could be a rhyolite. However, because rhyolite has such small crystals, it can be difficult to confidently identify without looking at the rock under a microscope or sending a sample of the rock for chemical analysis. If larger phenocryst crystals are present, this may help with identification. Rhyolite is commonly found with other volcanic rocks, such as obsidian and pumice. So, finding these rocks nearby is a good clue that a rock may be rhyolite.

FUN FACT: In Nevada there is a ghost town named Rhyolite, which is named after the rhyolite rock that is found in the vicinity of the town. Gold was discovered in this rock in the early 1900s, which led to a gold rush from 1905 to 1911. After 1920, the town was abandoned, but it became a tourist attraction and movie filming location. You can still visit this famous ghost town today.

BASALT

ROCK TYPE: Igneous

COLOR: Usually black or dark gray

MINERALOGY: Mostly plagioclase feldspar, olivine, and pyroxene

TEXTURE: Aphanitic, sometimes porphyritic or glassy

DESCRIPTION AND OCCURRENCE: Basalt is a fine-grained igneous rock that is usually black or dark gray in color. However, when basalt weathers it can turn a reddish brown or greenish brown color. Basalt is an extrusive rock, which means that it forms from lava that flows on Earth's surface. Because lava cools relatively quickly, basalt is composed of small crystals (since larger crystals do not have time to form), which are often too small to see with the naked eye. If you look at basalt under a microscope, you can see that these small crystals are mostly the minerals feldspar (plagioclase variety), olivine, and pyroxene. In addition, there may be small amounts of other minerals, such as magnetite. Sometimes, basalt may contain a few larger crystals (called "phenocrysts"), often of olivine (green or altered to reddish brown in color) or plagioclase (white in color). These larger crystals formed in the magma chamber of a volcano prior to the eruption of the magma as lava. When basalt contains no or very few phenocrysts, its texture is aphanitic. When basalt contains a significant number of phenocrysts, its texture is porphyritic.

Basalt sometimes has vesicles (cavities), which form when gases are released from the lava. These vesicles can become filled with secondary minerals, such as zeolite and calcite. The secondary minerals are often white in color, which can give the basaltic rock a spotted appearance (white spots on a dark rock). A vesicle that has been filled with a secondary mineral is called an amygdale. You can use the terms "vesicular basalt" and "amygdaloidal basalt" to describe basalt with these characteristics.

Basalt is a mafic rock, which means that it contains high amounts of the elements magnesium and iron and a low amount (relative to other rocks) of the element silicon. Basalt is related to another rock called gabbro, which has the same chemical composition but has much larger crystals. Basalt and gabbro are produced from the same type of melting, but gabbro is an intrusive rock (formed under the Earth's surface), which means that it forms from slow-cooling magma and there is time for larger crystals to form. Because basalt contains a significant quantity of heavy elements, such as iron, it is a relatively heavy rock, with a density of about 2.8 to 3.0 g/cm^3.

Basalt has a few different morphologies, depending on how and where it is erupted. When basaltic lava flows slowly across the Earth's surface, it often piles up into smooth coils that look like rope. This type of basalt is called "pahoehoe", a Hawaiian word that is pronounced "pah-hoy-hoy". When basaltic lava flows more quickly in open channels, its morphology is rougher and more blocky. This type of basalt is called "a'a," another Hawaiian word that is pronounced "ah-ah." When basalt erupts underwater, such as on the seafloor, it can form a type of basalt known as "pillow lava." Pillow lavas have a pillow-like shape and have a rim of glass, which forms when the basaltic lava cools rapidly when it comes into contact with water. Another basalt morphology is "columnar basalt," which sometimes forms when thick basalt flows cool slowly. This type of basalt consists of tall columns, which are usually hexagonal in shape.

Basalt is one of the most common rocks on Earth. The ocean floor consists mostly of basalt, which is erupted from underwater volcanoes. Sometimes, these underwater volcanoes erupt large volumes of basalt, which can pile up and rise above the water's surface and form islands. For example, this is what formed the Hawaiian Islands. Basalt can also erupt on Earth's continents. Several times in Earth's history, large amounts of basalt have erupted on land. These large eruptions are called continental flood basalts.

In the United States, basalt can be found in many places. However, one of the best places to see basalt is Hawaii, where you can even observe an actively erupting volcano! Another great place to see basalt is Craters of the Moon National Monument in Idaho, where you can see volcanic cones and lava flows. Thick basalt flows can be seen in the Colombia River Basalt, which is a continental flood basalt located in the states of Washington and Oregon. Two great places to see columnar basalt are Devils Tower National Monument in Wyoming and Devils Postpile National Monument in California.

IDENTIFICATION: If you see a dark, fairly heavy, fine-grained rock, then it is most likely basalt. If the basalt has weathered, you may need to break off a piece of the rock in order to view a fresh surface. Basalt should be fine-grained, with the crystals mostly too small to see with the naked eye. You may be able to observe some phenocrysts and/or amygdales. To confirm that a rock is a basalt, geologists generally use a hand lens or microscope to look for mafic minerals in the rock.

FUN FACT: Basalt is a common rock on the moon. Darker-looking areas of the moon, known as the "lunar maria," are composed of large flows of basalt. Lunar basalts are very old. Most erupted between 3 and 4 billion years ago!

PUMICE

ROCK TYPE: Igneous

COLOR: Usually white, light brown, or light gray

MINERALOGY: Mostly glass (mineraloid)

TEXTURE: Glassy, vesicular

DESCRIPTION AND OCCURRENCE: Pumice is a glassy, vesicular (full of vesicles or cavities) rock that forms at volcanoes when gas-rich magma erupts explosively and forms a frothy rock. The loss of the dissolved gases causes the magma to solidify rapidly. Therefore, pumice is mostly comprised of glassy material (a mineraloid), since there is no time for mineral crystals to form. Sometimes, pumice can contain small crystals that formed in the magma before the eruption.

Pumice most commonly forms at felsic (silica-rich) volcanoes, which tend to have higher gas contents. Felsic pumice can contain small crystals of minerals such as feldspar, amphibole, and pyroxene. Less commonly, pumice forms at mafic (silica-poor) volcanoes, particularly when they erupt underwater. Pumice is usually light in color, commonly white, light brown, or light gray. Mafic pumice may be darker in color. Pumice is an abrasive material and generally has a rough surface. Pumice is a very light rock with a density that is usually less than 1.0 g/cm^3, which means that it floats on water. In comparison, most igneous rocks have densities of about 2.5 to 3.0 g/cm^3.

Pumice can be found in abundance near volcanoes all over the world. Pumice found near volcanoes generally has a blocky or angular shape. It can be found in thick volcanic flows. Because pumice floats on water, it can also be found far away from the volcanoes that produced it, for example on beaches. Pumice that is found on beaches is often rounded because it has been weathered in the ocean and by waves in the beach environment.

Pumice is mined for use in construction and in the beauty industry. Pumice stones are commonly used for skin exfoliation. You may even have a pumice stone in your own bathroom! Pumice is also used as an abrasive material in some industrial processes, for example in the production of stone-washed jeans. In the United States, pumice is mined in several states, including Arizona, California, Idaho, Nevada, New Mexico, Oregon, and Washington. A great place to see pumice in the United States is Pumice Castle, a beautifully weathered outcrop of pumice that looks like a castle. Pumice Castle is located in Crater Lake National Park in Oregon and is part of a thick pumice layer that erupted from the Mount Mazama volcano approximately 70,000 years ago.

IDENTIFICATION: Pumice usually floats on water, so that is an easy way to identify it—drop it into a body of water (or a bowl of water) and see if it floats! No other rock floats on water. Scoria is a similar-looking rock, but it is denser than pumice and will not float on water.

FUN FACT: Pumice can float on water for years before it becomes waterlogged and sinks. Thus, pumice erupted from volcanoes can travel thousands of miles across oceans before washing up on distant beaches.

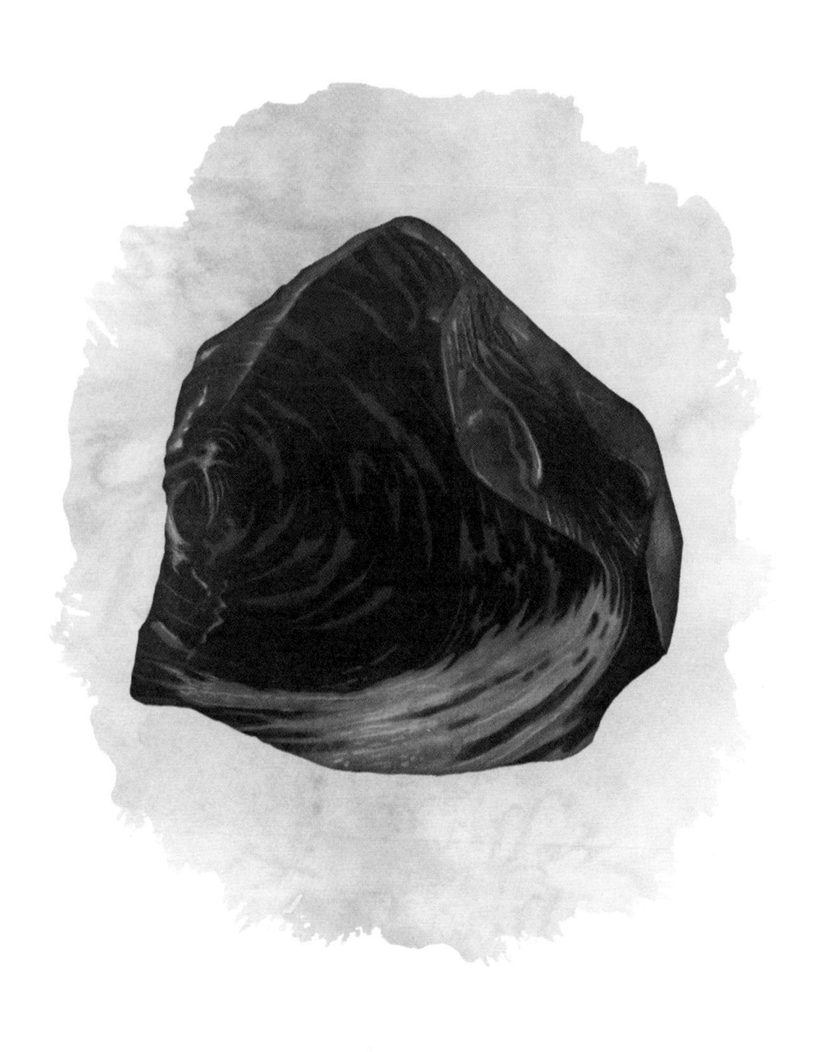

OBSIDIAN

ROCK TYPE: Igneous

COLOR: Usually black

MINERALOGY: Silica glass (mineraloid)

TEXTURE: Glassy, sometimes with small crystals

DESCRIPTION AND OCCURRENCE: Obsidian is a type of natural glass that is produced by some types of volcanoes. Obsidian forms when lava cools very quickly and there is no time for mineral crystals to form. Instead, the lava cools into an amorphous glass. Obsidian is silica-rich and forms in felsic (silica-rich) volcanoes. The reason that felsic volcanoes produce obsidian is that it is more difficult (it takes more time) for crystals to form in silica-rich lava. Obsidian is often found in association with silica-rich rhyolite, which is also a volcanic rock.

Obsidian is most commonly black in color. However, it can also be gray, brown, or green. Rarely, it can be other colors, such as red, yellow, orange, or blue. Obsidian is a natural glass, which is a mineraloid rather than a mineral (since a mineral must have a crystal structure, whereas glass has an amorphous structure).

Obsidian has a vitreous (glassy) luster and is about 5—6 on the Mohs Hardness Scale (similar in hardness to man-made glass). Obsidian has conchoidal fracture, which means that it fractures along curved surfaces. Obsidian can be translucent or opaque. One variety of obsidian, referred to as "snowflake obsidian," contains small white crystals of cristobalite, a silica-rich mineral. The cristobalite crystals form either as a result of devitrification (changing of glass to a mineral, which can occur over time) or by deposition from volcanic gases.

In the United States, obsidian can be found near volcanoes in several western states, including Oregon, Washington, California, Wyoming, Montana, Arizona, and New Mexico. A famous obsidian location is Obsidian Cliff, which is located in Wyoming in Yellowstone National Park. This cliff consists of 150—200 feet of obsidian and has been designated a National Historic Landmark.

IDENTIFICATION: Obsidian generally looks like shiny black glass and often has curved surfaces from its characteristic conchoidal fracture. You need to be careful not to confuse obsidian with man-made glass or slag, which can look similar. If you find glassy material outside, it is important to understand if volcanoes are nearby. Obsidian will only be found near volcanoes, which may be dormant or extinct rather than actively erupting. However, obsidian is generally found near fairly young (by geologic standards, anyway!) volcanoes, because glass is destroyed over time through weathering processes. Thus, it is rare to find obsidian that is more than a few million years old.

FUN FACT: Obsidian has been used by people for thousands of years in weapons, tools, and ornaments. Obsidian is a very useful material because the conchoidal fracture of the glass can produce very sharp edges.

PERIDOTITE

ROCK TYPE: Igneous and metamorphic

COLOR: Green, black, or brown

MINERALOGY: Mostly olivine and pyroxene

TEXTURE: Coarse-grained phaneritic

DESCRIPTION AND OCCURRENCE: Peridotite is comprised of olivine and pyroxene, both orthopyroxene and clinopyroxene, and can form as an intrusive (formed underground from magma) igneous rock. Peridotites are classified by the relative proportions of the minerals olivine, orthopyroxene, and clinopyroxene that are present in the rock. Dunite is a type of peridotite that consists mostly of the mineral olivine (90% or more). Harzburgite is a type of peridotite that contains relatively little clinopyroxene (5% or less). In contrast, wehrlite is a type of peridotite that contains relatively little orthopyroxene (5% or less). Lherzolite is a type of peridotite that contains olivine as well as a mix of both clinopyroxene and orthopyroxene. Although peridotite is commonly thought of as an igneous rock, peridotite rocks often represent portions of the Earth's mantle layer (found beneath the crust), which did not crystallize from magma. Therefore, mantle peridotites are better classified as metamorphic rocks.

Fresh peridotite is usually green or black in color, with pale green olivine and dark green or black pyroxene. In addition to olivine and pyroxene, peridotite commonly contains other minerals, such as hornblende, chromite, garnet, plagioclase feldspar, spinel, mica, and magnetite. Peridotite is easily altered and will often look brown or reddish brown when weathered, due to the transformation of the olivine to other minerals that have oxidized iron. Alteration of peridotite can also lead to the formation of green-colored serpentine minerals, as well as white brucite, talc, and carbonate minerals. When a peridotite is very highly altered, it can turn into a related rock called serpentinite.

Peridotite is an ultramafic rock, which means that it contains a high amount of magnesium and iron and relatively little silica (compared to other igneous rocks). Due to the high density of the iron and magnesium minerals, peridotite is a heavy rock, with a density of approximately 3.2 to 3.3 g/cm^3.

If you look at the Earth as a whole, peridotite is one of the most common rocks that makes up the planet. This is because the upper part of Earth's mantle is composed mostly of peridotite. However, peridotite is harder to find on Earth's surface, where it comprises only approximately 1% of rocks found. The peridotite that is found at Earth's surface usually represents part of the Earth's mantle that has been brought to the surface through the movement of tectonic plates. Large exposures of peridotite are found in ophiolites, which are sections of ocean crust and underlying mantle that are exposed on land by obduction (placement of oceanic crust on top of other crust), usually when an ocean basin closes. Small pieces of peridotite are also found at the Earth's surface as xenoliths (literally "foreign rocks") in volcanic rocks, such as basalts and kimberlites. Peridotite can also form when olivine and pyroxene crystals accumulate in a magma body, which later cools and forms a rock. For example, this occurs in some large igneous intrusions.

In the United States, great places to see peridotite are in ophiolite rocks found in Oregon, Washington, and California. Good places to find peridotite xenoliths are the San Carlos Volcanic Field in Arizona and Kilbourne Hole in New Mexico.

IDENTIFICATION: Peridotite can be a difficult rock to identify, since it is often highly altered. If the peridotite is fresh, you can look for coarse crystals of light green olivine and dark green to black pyroxene. If the peridotite is altered, the rock may turn a brown or brownish red color, and it can be difficult to confidently identify the crystals. Serpentine and other alteration minerals may also be present. Peridotite rocks are heavy, so heaviness is a good clue that a rock could be a peridotite. The presence of related ophiolite rocks, such as pillow basalts or marine sediments, nearby is another good clue that a rock could be a peridotite. Geologists often identify and classify peridotite rocks using microscopes and chemical analyses.

FUN FACT: The gemstone peridot is found in peridotite. Peridot is a light green, translucent variety of the mineral olivine. In the United States, this gem is mined near the aptly named town of Peridot, Arizona.

PEGMATITE

ROCK TYPE: Igneous

COLOR: Variable

MINERALOGY: Variable

TEXTURE: Very coarse-grained phaneritic

DESCRIPTION AND OCCURRENCE: Pegmatite is a term that is used to refer to an igneous rock that has very large crystals. Pegmatites generally have crystals that are over 1 centimeter in size and frequently have much larger crystals. Some pegmatites have crystals that are over 10 meters in size!

Pegmatites are variable in color, mineralogy, and chemical composition. This is because pegmatite is a textural term, so it can be used to refer to any igneous rock with large crystals. Geologists commonly use the term pegmatite along with another term that describes the composition of the rock. For example, a geologist may describe a rock as a "granite pegmatite" or a "gabbro pegmatite." Granite pegmatites are the most common type of pegmatite and are comprised mostly of alkali feldspar, quartz, and mica. They can also contain a wide variety of other minerals, such as tourmaline, beryl, topaz, lepidolite, spodumene, fluorite, magnetite, zircon, chalcopyrite, corundum, and columbite.

The formation of pegmatites is not completely understood by geologists, and there is still much to learn about the details of how different types of pegmatites form. However, many pegmatites probably form from the last part of a magma body that crystallizes. As igneous rocks crystallize from magma, the remaining magma tends to become more and more enriched in fluid, which contains a significant quantity of volatiles, such as water, carbon dioxide, fluorine, and chlorine. The volatile-rich fluid enables the growth of very large crystals. The magmas that form pegmatites are often enriched in elements that are not easily incorporated into the structure of regular igneous rocks. Because of this, pegmatites are sometimes mined for rare elements, such as lithium, tin, beryllium, boron, niobium, zircon, uranium, thorium, and tantalum. Due to their unusual chemistry, pegmatites often contain rare minerals, including gemstones. Pegmatites are an important source of the gemstones emerald, aquamarine, tourmaline, and topaz.

Pegmatites are often found near other igneous rocks. For example, granite pegmatites are commonly found near granites. In the United States, a great place to see pegmatite is the Black Hills region of South Dakota, for example in Wind Cave National Park. Another great place to see pegmatite is in Black Canyon of the Gunnison National Park in Colorado, where pegmatite can be seen throughout the 48-mile-long canyon. Pegmatites are also commonly found in the White Mountains of New Hampshire and Maine, as well as in mountainous regions of western North Carolina, northern Georgia, and southern California.

IDENTIFICATION: If you see a rock with very large crystals that are over 1 centimeter in size, then that rock is most likely a pegmatite. Since the crystals are large, they should be fairly easy to identify. Look for quartz, feldspar, and mica, since these minerals are commonly found in pegmatites.

FUN FACT: Pegmatite veins can be seen at Mount Rushmore National Memorial in South Dakota. The giant sculptures of the heads of the presidents were carved into granite rock, which contains veins of granite pegmatite. The pegmatites can be seen as light-colored veins on the foreheads of the sculptures.

SANDSTONE

ROCK TYPE: Sedimentary

COLOR: Variable—often white, tan, brown, gray, pink, or red

MINERALOGY: Mostly quartz, often contains feldspar and mica

TEXTURE: Clastic

DESCRIPTION AND OCCURRENCE: Sandstone is one of the most common sedimentary rocks. It is comprised of sand-sized (0.0625 to 2 millimeter size) grains in a finer-grained matrix. The mineralogy of sandstone varies, but it most commonly consists primarily of quartz and feldspar. However, some sandstones are composed of carbonate minerals or even evaporite minerals, such as gypsum. As long as the grains are sand-sized, a sedimentary rock can be classified as a sandstone.

Sandstones are most commonly classified by the proportions of quartz, feldspar, and rock fragments (lithics) they contain. The terms "quartzarenite" (quartz-rich), "arkose" (feldspar-rich), and "litharenite" (lithics-rich) describe the three end member compositions. There are other classifications for sandstones that describe the rounding or sorting of the sand grains or the composition of the finer-grained matrix. For example, the term "greywacke" is sometimes used to describe a sandstone with a significant fraction of clay in the matrix. When quartz-rich sandstone is metamorphosed, it can turn into a rock called quartzite.

Sandstone color is highly variable, depending on its mineral composition and degree of alteration. Quartz-rich sandstones tend to be white or tan in color. When iron oxides are present, sandstone can be a pink or red color. Sandstones can also be brown, gray, black, and yellow.

Sandstones often contain layers known as bedding. When the layers are formed at different angles at different times, sandstones can contain a feature known as cross-bedding. For example, cross-bedding can form in sand dunes when wind or water currents cause the dunes to partially migrate over time. Sand grains are pushed over the tops of the dunes and redeposited at angle on the down current side of the dunes. Sand dunes with cross-bedding can later be transformed into sandstone rock.

Sandstone is formed where sand-sized particles accumulate, such as deserts, riverbeds, and beaches. In the United States, one of the most extensive sandstone formations is the Navajo Sandstone, which stretches across the states of Utah, Nevada, Arizona, and Colorado. The Navajo Sandstone forms beautiful cliffs and other rock features that can be seen in places such as Zion National Park and Canyonlands National Park in Utah and Red Rock Canyon National Conservation Area in Nevada.

IDENTIFICATION: Sandstone is fairly easy to identify. Look for a rock made up of sand-sized particles. If you touch sandstone, it will feel gritty (like sandpaper). The distinctive cross-bedding that is present in many sandstones is a good way to identify sandstone from a distance, although not all sandstone has cross-bedding.

FUN FACT: Delicate Arch, an iconic 16-meter-tall rock arch in Arches National Park in Utah, consists of sandstone rock. Specifically, Delicate Arch formed in the Entrada Sandstone, which is approximately 140 to 180 million years old. Weathering of the sandstone over time left the rock arch behind. This famous sandstone rock arch is commonly depicted on Utah license plates.

MUDSTONE

ROCK TYPE: Sedimentary

COLOR: Variable—often brown, gray, black, red, white, or green

MINERALOGY: Variable—often quartz, feldspar, mica, and clay minerals

TEXTURE: Clastic

DESCRIPTION AND OCCURRENCE: Mudstone is a sedimentary rock that is comprised of a mixture of clay-sized (less than 2 microns) and silt-sized (2 micron to 63 micron) grains. The terms "claystone" and "siltstone" are used for mudstone rocks that are composed dominantly of either clay-sized or silt-sized grains. "Shale" is a common term used for a mudstone that is fissile, meaning that it easily breaks apart in layers.

Mudstone can be many colors, including brown, gray, black, red, white, and green. The mineralogy of mudstone is variable. However, it commonly contains quartz, feldspar, and mica as well as clay minerals, such as kaolinite, smectite, and illite. Mudstone also commonly contains iron oxide minerals, such as hematite. If hematite is present, it can give the mudstone a reddish color. Mudstone can also contain grains of golden-colored pyrite. Mudstone commonly contains fossils, which are variable depending upon the age of the rock and the animals that were alive when the rock was formed.

Mudstone generally forms in quiet waters, where small clay- and silt-sized particles can settle at the bottom. Mudstone is commonly formed in lakes and in the ocean. Over geologic time, these water bodies can disappear (for example due to climate change or the closing of an ocean due to plate tectonic movements), leaving mudstone rock behind. Thus, mudstone can be found in many environments, including in modern-day deserts and mountains.

In the United States, a good place to find marine mudstone is in the Mancos Shale, which was formed in a large inland sea that covered a large portion of the western United States approximately 80 to 95 million years ago. The Mancos Shale can be seen in the states of Arizona, Colorado, New Mexico, Utah, and Wyoming. Another good place to see mudstone is in the Morrison Formation, which can be found in many states, with large outcrops in Wyoming, Colorado, and Utah. The Morrison Formation is approximately 150 million years old and also contains other sedimentary rocks, such as sandstone and limestone. The sandstone units of the Morrison Formation are famous for containing dinosaur bones, such as the ones on display at Dinosaur National Monument. Sometimes, the mudstone units preserve dinosaur footprints, for example those at Purgatoire River Tracksite in Colorado.

IDENTIFICATION: If you see a fine-grained rock that is fairly uniform in color, it could be mudstone. Geologists often lick (yes, lick!) a rock to tell if it contains silt-sized particles to determine if the mudstone is mostly silt or clay. A rock that contains silt-sized particles will feel gritty on the tongue and will indicate that the rock is a siltsone, which is a coarse-grained mudstone. In contrast, a rock that contains only clay-sized particles will feel smooth on the tongue and will be claystone, which is a fine-grained mudstone. Geologists often use microscopes to study mudstones in order to better see the small grains that form these rocks.

FUN FACT: Study of mudstones on Mars has provided information on the presence of water and organic molecules on the planet. Analysis of 3-billion-year-old mudstones in Gale Crater by the Curiosity Rover indicates that there was once a freshwater lake in the crater, which became salty as it evaporated and eventually disappeared. Analysis by the Curiosity Rover also showed that the Gale Crater mudstones contain organic molecules. The presence of both water and organic molecules in the mudstones makes scientists wonder if life could have existed in the crater in the past, although no firm evidence of life on Mars has been found to date.

LIMESTONE

ROCK TYPE: Sedimentary

COLOR: Variable—often white or gray

MINERALOGY: Mostly calcite or aragonite

TEXTURE: Clastic or non-clastic

DESCRIPTION AND OCCURRENCE: Limestone is a common sedimentary rock that is composed mostly of calcium carbonate, in the form of either the mineral calcite or the mineral aragonite. Aragonite and calcite have the same chemical composition but have different mineral structures. Aragonite is often made through biological processes. A related rock called dolostone contains a significant amount of the mineral dolomite, which is a carbonate mineral that contains both calcium and magnesium. Limestone does not contain a significant amount of dolomite. There are two general categories of limestone: biological limestone and chemical limestone.

Biological limestones form from the accumulation of sediments formed from the carbonate skeletons and shells of marine organisms, such as plankton, mollusks, and corals. Coquina is a type of biological limestone that consists mostly of shell fragments or whole shells. Chalk is a type of biological limestone that is fine-grained and composed mostly of the shells of plankton, such as foraminifera or coccolithophores. Biological limestones form in lacustrine (lake) or marine environments, most often in shallow waters. They can later be exposed on land, for example when lakes dry up, sea level changes, or when tectonic plate movement uplifts marine rocks. Sometimes, the fossils present in biological limestone are obvious to see with the naked eye. Other times, they are too small to see with the naked eye, or they may have been broken up or recrystallized.

Chemical limestones form through precipitation from either salt water or fresh water. For example, stalactites and stalagmites are a type of limestone that precipitates from water in caves. Travertine is a type of limestone that precipitates from springs, particularly hot springs. Tufa is a type of limestone that precipitates from alkaline waters, for example in a lake.

Limestone can be found in many places around the world. The White Cliffs of Dover in England, which are composed mostly of chalk, are an example of a famous limestone formation. In the United States, a great place to see chalk is the Niobrara Chalk in western Kansas. This chalk contains many well-preserved marine fossils. Another great place to see limestone is in a cave, where you can see limestone in the stalactites and stalagmites. Some good caves to visit are Howe Caverns in New York, Mammoth Cave in Kentucky, Wind Cave in South Dakota, Bridal Cave in Missouri, and Carlsbad Caverns in New Mexico. A great place to see travertine is Yellowstone National Park in Wyoming, and a great place to see tufa is Mono Lake in California.

IDENTIFICATION: Limestone is a soft rock, since it is made of soft carbonate minerals (hardness of about 3). Limestone is most commonly a light color, such as white or gray. However, bear in mind that limestone also comes in many other colors, such as yellow and brown. A good way to identify limestone is to test if it fizzes when it is exposed to a small amount of dilute acid, such as vinegar. However, be careful with this test. Other carbonate rocks, such as marble, will also fizz when exposed to acid. Dolostone will also react with acid, but it will react less vigorously than limestone. Biological limestones will often contain fossils or fossil fragments, for example pieces of shell and coral.

FUN FACT: The Lincoln Memorial in Washington, D.C., was constructed using limestone from Illinois, the home state of Abraham Lincoln. This limestone was used to make the columns and interior walls of the monument. Other materials used in the monument's construction include granite from Massachusetts and marbles from Colorado, Georgia, Tennessee, and Alabama.

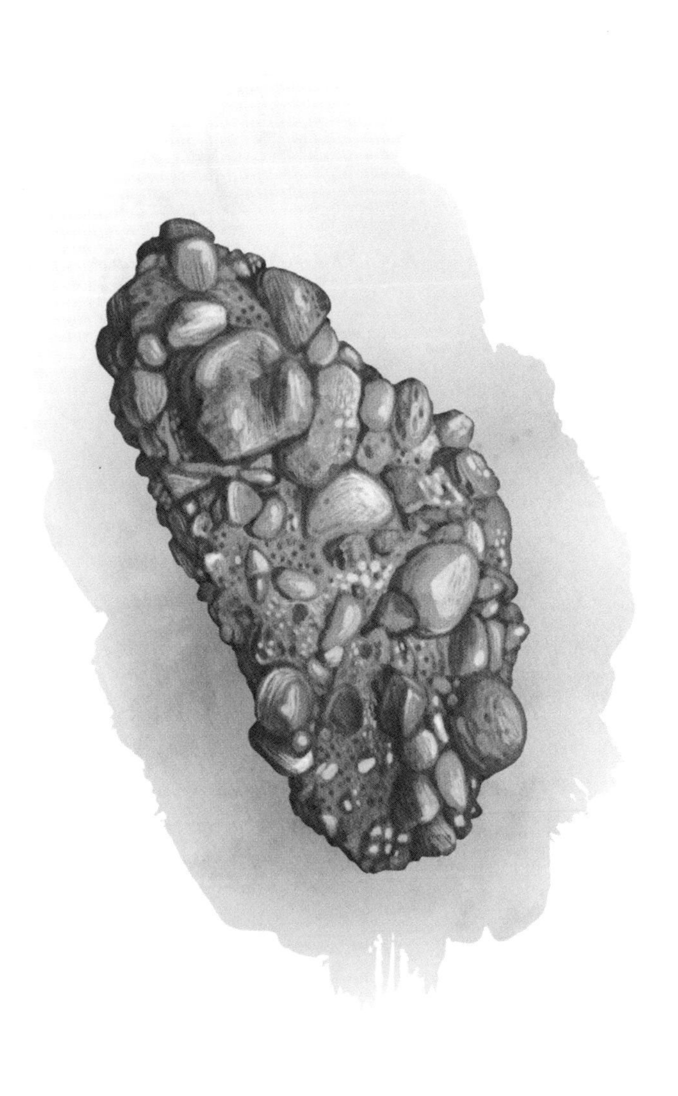

CONGLOMERATE

ROCK TYPE: Sedimentary

COLOR: Variable

MINERALOGY: Variable

TEXTURE: Clastic

DESCRIPTION AND OCCURRENCE: Conglomerate is a sedimentary rock that is comprised of rounded clasts (or fragments) of rocks, which are cemented together with finer-grained material, such as sand or silt. In order for a rock to be called a conglomerate, the clasts must be at least 2 mm in diameter. Conglomerates often contain larger clasts that are several centimeters in diameter. Some conglomerate formations are made out of large boulders. In addition, in order for a rock to be called a conglomerate, the clasts must make up at least 30% of the rock.

Because conglomerates are made up of fragments of other rocks, they are highly variable in color, mineralogy, and chemical composition. However, it is common to find highly durable rocks, such as those that are rich in quartz, in conglomerates. This is because durable rocks are more likely to survive the transport and weathering processes that create large, rounded clasts. The finer-grained cementing material often contains a significant amount of quartz or calcite. If the finer-grained material is bright white in color, then it is probably composed mostly of calcite.

Conglomerates form from gravels that are cemented together. Thus, conglomerates will be found in places where gravel accumulates, such as rivers, beaches, or marine environments. Good places to look for conglomerates are riverbeds and beaches. Conglomerates can also form when rock material is transported and deposited by glaciers.

IDENTIFICATION: Conglomerates are easy to identify. Look for a rock comprised of rounded rock fragments that are at least 2 mm in diameter and cemented together with finer-grained material. If the rock fragments are angular instead of rounded, then the rock is a breccia rather than a conglomerate. Remember that conglomerates can be comprised of many different types of rocks and thus can have many different colors and textures. The key to identifying a conglomerate is noting the shape and size of the rock fragments.

FUN FACT: Some conglomerates are called "puddingstones." This informal term is used to describe conglomerates that have rounded clasts that have a strong color contrast with the finer-grained matrix, often darker clasts in a light matrix. These rocks look similar to plum pudding, which has dark raisins (or other dried fruit) in lighter-colored cake.

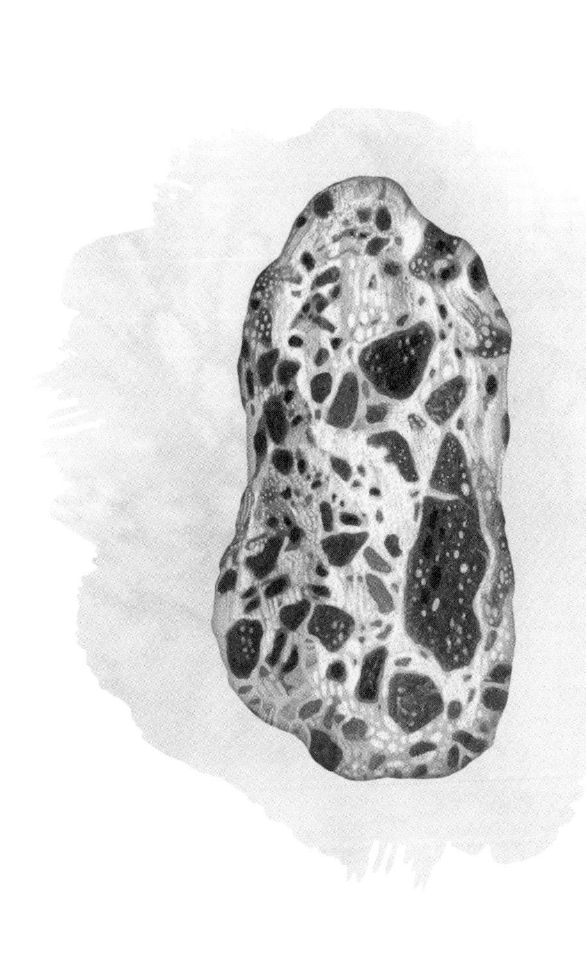

BRECCIA

ROCK TYPE: Sedimentary

COLOR: Variable

MINERALOGY: Variable

TEXTURE: Clastic

DESCRIPTION AND OCCURRENCE: Breccia is a sedimentary rock that is comprised of angular clasts (or fragments) of rocks, which are cemented together with finer-grained material, such as sand or silt. Similar to a conglomerate, the clasts must be at least 2 mm in diameter and must comprise at least 30% of the rock. However, the clasts in a breccia must be angular in shape whereas the clasts in a conglomerate are rounded. The clasts can range up to centimeters or meters in size, and it is common for breccias to contain clasts of different sizes.

Because breccias are made up of fragments of other rocks, they are highly variable in color, mineralogy, and chemical composition. The clasts can be any type of igneous, sedimentary, or metamorphic rock. The finer-grained cementing material often contains a significant amount of quartz or calcite.

Breccia can form in many different ways. One way that breccia can form is when rock fragments accumulate at the bottom of a slope or cliff and later become cemented together. Breccia can also form when a rock body collapses, for example in a cave or sinkhole. Additionally, movement along geological faults can create breccia. In this breccia, you can sometimes see offset fragments of a rock that was split in two by the fault movement. Volcanic activity can also create breccia, when an explosive eruption breaks up rock or when part of a volcanic crater collapses. Flow of hydrothermal (hot) fluids through the Earth's crust can also create breccia. Another way that breccia can form is through meteorite impacts.

Because breccia forms in so many different ways, you can see it in many different places. In the United States, one great place to see breccia is Mosaic Canyon in Death Valley National Park in California. The canyon was named for the beautiful breccias, resembling mosaics, that were formed in the canyon by geological faulting.

IDENTIFICATION: If a rock is comprised of angular fragments, then it is most likely a breccia. Check that the fragments are at least 2 mm in diameter and that they are cemented together by finer-grained material. If the fragments are rounded instead of angular, then the rock is a conglomerate rather than a breccia.

FUN FACT: When Apollo 16 astronauts visited the Descartes Highlands region of the moon in 1972, they thought that they would find volcanic rocks. Instead, most of the rocks that they observed and collected turned out to be breccias. These breccias were formed by meteorite impacts on the surface of the moon.

TILLITE

ROCK TYPE: Sedimentary

COLOR: Usually gray, brown, or green

MINERALOGY: Variable

TEXTURE: Clastic

DESCRIPTION AND OCCURRENCE: Tillite is a word used to describe a rock that forms from the consolidation of till, which is a type of sediment that is deposited by a glacier. Till is composed of poorly sorted sedimentary material and contains grains ranging from clay-size (less than two microns) all the way up to boulders that can be tens of meters in size. In sediments, it is common for grains to be sorted by size during the transport and depositional processes and for there to be layers of grains of similar size. However, till is very poorly sorted and can have clay-sized particles next to enormous boulders. This is because glaciers pick up and transport sediments of all different sizes. Eventually, these sediments are deposited as till at the edge of a glacier. Later, the till can be consolidated into tillite rock. Tillite can contain fragments of all different rock types. Glaciers can move rock fragments very far. The rock fragments found in tillite are sometimes traced to bedrock sources that are hundreds of miles away.

The color, mineralogy, and chemical composition of tillite are variable. However, tillite commonly contains a significant amount of clay that is gray, brown, or green in color. Within this clay you can find sand particles, as well as larger rock clasts, which can be angular or rounded. The angular clasts often represent pieces of bedrock that were ripped up by the glacier. The rounded clasts may originate from riverbeds that the glacier passed over.

Although many poorly sorted sedimentary rocks were deposited by glaciers, this is not the only geologic environment in which these rocks form. For example, submarine landslides and mudflows caused by volcanic eruptions can also create poorly sorted sedimentary rocks. The term tillite is only used for rocks that originated from a glacier. For this reason, geologists often use the term "diamictite" when they are first describing these types of sedimentary rocks. Diamictite is a descriptive term that can be used to classify any poorly sorted sedimentary rock. When geologists are confident that a diamictite originated from a glacier, then they call this rock a tillite.

Tillite can be found in the vicinity of current glaciers, as well as in places where glaciers existed in the past. In the United States, you can find tillite in northern states, where the Laurentide Ice Sheet existed in the past. This ice sheet advanced and retreated across the northern part of the United States several times during the past 2.5 million years. This ice sheet was up to two miles thick and left tillite in several states, including Alaska, Minnesota, Wisconsin, Michigan,

New York, and the New England states. The Laurentide Ice Sheet also carved out the depressions that became the five Great Lakes.

IDENTIFICATION: Tillite is easy to identify. Look for a rock that has particles of many different sizes and which has little to no sorting of these particles. If you are in an area that had glaciers in the past, you may also be able to spot something called a glacial erratic. This is a boulder that was left behind by a glacier. Glacial erratics are usually found by themselves with no other boulders nearby. Glacial erratics can be very large—some are as large as houses!

FUN FACT: Geologists sometimes sample tillite (or till) when they are looking for rare minerals, such as diamonds and gold. If they find these minerals (or related minerals, which are called "indicator minerals") in the tillite, they can often track down the bedrock source of the minerals by mapping out the travel path of the glacier that deposited the tillite.

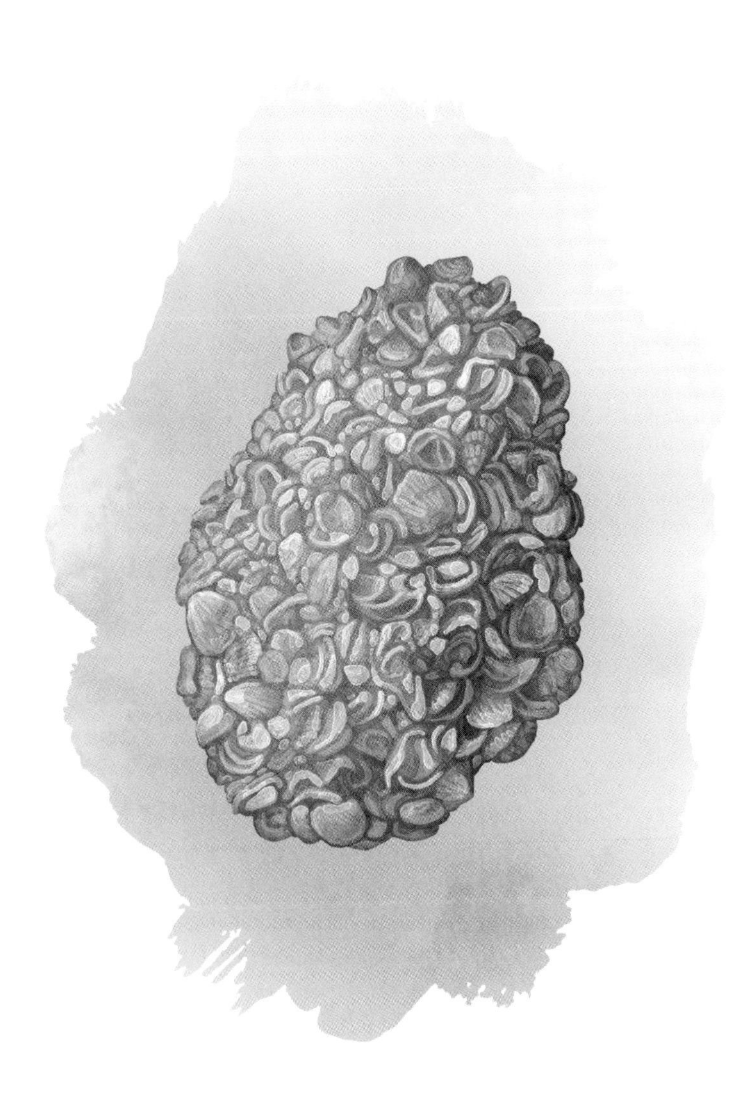

COQUINA

ROCK TYPE: Sedimentary

COLOR: Generally white or tan

MINERALOGY: Mostly calcite and/or aragonite

TEXTURE: Clastic

DESCRIPTION AND OCCURRENCE: Coquina is a clastic sedimentary rock that is entirely (or nearly entirely) composed of whole shells and/or shell fragments. In order for a rock to be called coquina, the shells or shell fragments must be on average 2 mm or greater in size. If they are smaller, then the rock is called microcoquina. In addition, in order for a rock to be called coquina the shells or shell fragments should be lightly cemented together. If they are firmly cemented together into a harder, denser rock, then the rock is called coquinite. Because coquina is comprised of shells that are made of calcite and/or aragonite, it is technically a limestone. Most coquina is made from marine shells. However, coquina made from freshwater shells also exists.

Coquina rock typically forms in high-energy coastal environments, such as beaches, sand bars, and tidal channels. In these environments, there is a supply of shell material, but finer clay- and silt-sized particles can be winnowed away. This enables the larger shells and shell fragments to be concentrated and cemented into coquina rock. Coquina can be found in coastal environments around the globe. In the United States, large deposits of marine coquina are found in Florida and North Carolina. Freshwater coquina can be found in paleo-lake environments in Wyoming.

IDENTIFICATION: Coquina is easy to identify—it looks like a pile of shells that have been cemented together. The rock is generally soft and easy to crumble in your hands. You are most likely to find coquina at a beach.

FUN FACT: In the 1600s, coquina stone was quarried and used to build the walls of Castillo de San Marcos, a Spanish fort in St. Augustine, Florida. Amazingly, this fort made out of seashells has survived for more than 300 years!

SLATE

ROCK TYPE: Metamorphic

COLOR: Usually gray

MINERALOGY: Mostly clay minerals, mica, and quartz

TEXTURE: Fine-grained and foliated

DESCRIPTION AND OCCURRENCE: Slate is a fine-grained metamorphic rock that is formed through low pressure and low temperature metamorphism of a sedimentary rock. The precursor rock is generally a clay-rich sedimentary rock, such as a mudstone. A key characteristic of slate is that it easily splits into layers along foliation planes (also called cleavage planes), which form during the metamorphism and which are generally perpendicular to the original layering in the sedimentary rock.

The mineral crystals in slate are too small to be seen with the naked eye or even with a hand lens. However, under a microscope it can be seen that slate is mostly comprised of clay minerals, mica, and quartz. Slate may also contain small amounts of other minerals, such as feldspar, chlorite, hematite, and pyrite. Sometimes, the pyrite crystals are large enough to see without magnification.

Slate is most commonly light gray or dark gray in color. Slate can also be brown, black, green, red, or purple in color. The color of slate is related to its composition. Dark-colored slates generally contain organic matter. Green slates are rich in chlorite, while red slates are rich in hematite.

Slates form through low-grade regional metamorphism of sedimentary rocks, for example when a sedimentary basin is subjected to a change in tectonic forces when two tectonic plates converge. Slate is a common sedimentary rock and can be seen many places. Slate is commonly used as a building stone, particularly for roof tiles. So, a great place to look for slate is on the roofs of buildings.

In the United States, a good place to see slate is the Slate Belt in Pennsylvania, where slate has been quarried since the 1700s. Another great place to see slate is Slate Valley in upstate New York, along the border with Vermont. You can even visit the Slate Valley Museum in Granville, New York. The museum has exhibits on the history of slate quarrying in the region.

IDENTIFICATION: If you see a fine-grained rock that is uniform in color, it could be slate. Slate can look similar to mudstone and other fine-grained sedimentary rocks, but slate will break into flat slabs. Although slate can occur in a variety of colors, it is most commonly light or dark gray.

FUN FACT: The terms "clean slate" and "blank slate" originated in the 1800s from the use of slate rock for writing in schools, as well as to keep track of debts.

SCHIST

ROCK TYPE: Metamorphic

COLOR: Usually gray or silvery gray, sometimes green, blue, or black

MINERALOGY: Variable

TEXTURE: Medium- to coarse-grained, foliated

DESCRIPTION AND OCCURRENCE: Schist is a metamorphic rock that is formed under medium-grade pressure and temperature conditions. Schist is formed through higher grade metamorphism than slate but lower grade metamorphism than gneiss. Schists have thin foliations (layers) that are formed by platy (elongated) minerals, such as mica and chlorite. Schists contain 50% or more platy minerals and easily split apart into thin layers.

Schist can form from a variety of different precursor rocks and thus is variable in composition. Mica (both muscovite and biotite) is commonly found in schist, which makes schist shiny and sparkly. Some schists will contain chlorite, a shiny green mineral, rather than mica. Schist can also contain many other minerals, such as garnet, kyanite, amphibole, talc, glaucophane, and graphite. Schists are often named by the dominant minerals that are found in the schists. For example, a geologist might refer to a rock as a "kyanite garnet mica schist" or a "talc chlorite schist." Most schists are gray or silvery gray in color, sometimes with grains of different colored minerals, such as red garnet or blue kyanite. Chlorite schists are green in color while glaucophane schists are bluish gray in color. Graphitic schist, which forms through the metamorphism of organic-rich sedimentary rock, contains graphite and is black in color.

Schist generally forms through regional metamorphism, for example when two tectonic plates converge. Schist is a common metamorphic rock that can be found in many places. You are likely to encounter schist in mountainous areas, where schist is commonly found alongside other metamorphic rocks, such as gneiss.

In the United States, you can find schist in mountainous regions, such as the Appalachian Mountains, the Adirondack Mountains, and the Rocky Mountains. One fantastic location to see schist is the Grand Canyon in Arizona. The lowermost geologic layer of the canyon contains 1.7-billion-year-old schist. You can see this rock from the canyon rim. If you are up for a challenge, you can hike to the bottom of the canyon to see it up close.

IDENTIFICATION: If you see a shiny, sparkly rock that easily splits apart into layers, then it is likely a schist. A significant proportion of the minerals should be elongated, leading to strong foliation (thin layering of the rock). Look for mica (either silver muscovite or black biotite) or green chlorite, which are commonly found in schist. If you spot other minerals, such as garnet, then you may be able to identify the rock using a more specific term, such as "garnet mica schist." You should be able to see the minerals with the naked eye, but a hand lens can help you identify them more confidently.

FUN FACT: The word "schist" originates from a Latin phrase "lapis schistos," which means "stone that splits easily." This term was used by the famous ancient Roman naturalist Pliny the Elder.

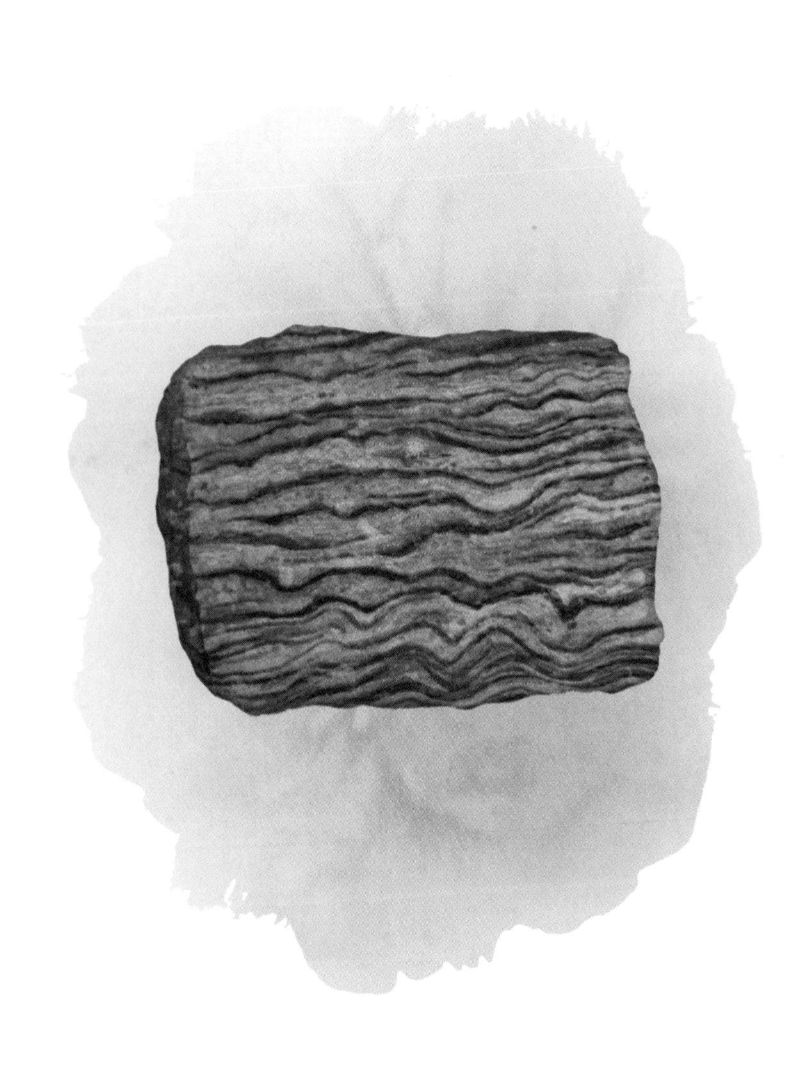

GNEISS

ROCK TYPE: Metamorphic

COLOR: Bands of alternating light and dark colors

MINERALOGY: Variable

TEXTURE: Medium- to coarse-grained, foliated

DESCRIPTION AND OCCURRENCE: Gneiss (pronounced like "nice") is a metamorphic rock that is formed by high pressure and high temperature transformation of either igneous or sedimentary rock. Gneiss is a term used to describe a metamorphic rock that has compositional layering. The layers (or bands) must be at least 5 mm thick and usually have alternating light and dark colors. The light layers are generally white or pink in color, while the dark layers are black or brown in color. Unlike slate and schist, gneiss will not split apart into thin layers.

Gneiss is variable in composition, depending upon the precursor rock that was metamorphosed. A gneiss that formed through the metamorphism of an igneous rock is called an orthogneiss. A gneiss that formed through the metamorphism of a sedimentary rock is called a paragneiss. The mineralogy of gneiss is variable. However, the light layers will consist of felsic (silica-rich) minerals while the dark layers will consist of mafic (silica-poor) minerals. Quartz and feldspar are commonly found in the light layers. Pyroxene, hornblende, and mica (the dark biotite variety) are commonly found in the dark layers. Some gneiss will contain many red garnet crystals. Gneiss also sometimes contains large crystals (called "porphyroblasts"), usually of feldspar, that are surrounded by the light and dark layers. This type of gneiss is called "augen gneiss" from the German word for eyes ("augen"), since the large crystals resemble eyes.

Gneiss generally forms through regional metamorphism, for example when two tectonic plates collide and create mountains. The cores of mountains are often composed of gneiss. In addition, gneiss makes up a significant portion of the stable interior portions of the continents, although most of this gneiss lies underneath other rocks. Many of the gneisses found in the stable parts of continents are very old. Sometimes, uplift and erosion expose gneiss at Earth's surface.

In the United States, a great place to see gneiss is Grand Teton National Park in Wyoming, where the mountains are largely composed of gneiss. For example, gneiss can be seen along the trail to Inspiration Point, a popular hike in the park. Another good place to see gneiss is the Minnesota River Valley, where you can see the 3.5-billion-year-old Morton Gneiss, some of the oldest rock in the world. You can see this gneiss near the towns of Morton and Granite Falls.

IDENTIFICATION: Gneiss is easy to recognize by its alternating layers of light and dark minerals, which look like zebra stripes. The layers sometimes look curved or wavy, from folding of the rock during metamorphism.

FUN FACT: Gneisses are the oldest rocks on Earth, excluding meteorites. The Acasta Gneiss, which is located in the Northwest Territories in Canada, is more than four billion years old!

MARBLE

ROCK TYPE: Metamorphic

COLOR: Variable—often white

MINERALOGY: Mostly calcite or dolomite

TEXTURE: Medium- to coarse-grained, usually non-foliated

DESCRIPTION AND OCCURRENCE: Marble is a rock that forms through the metamorphism of carbonate-rich sedimentary rocks, such as limestone and dolostone. Marble is generally composed mostly of calcite or dolomite. When a marble is rich in dolomite, it is called a "dolomitic marble." Marble can contain small amounts of additional minerals, such as quartz, feldspar, mica, pyrite, and many others.

Marble forms when limestone or dolostone is metamorphosed, either through regional metamorphism (caused by a regional increase in temperature and pressure conditions, for example due to burial underneath other rocks) or through contact metamorphism (caused by intrusion of igneous rocks). The original carbonate minerals in limestone or dolostone are recrystallized, making them larger and interlocking. The calcite and dolomite crystals in marble are often shiny, which makes the rock sparkle.

Marble is often white. However, it can be many colors, including pink, yellow, and brown. Most marbles are light in color. Some marbles are uniform in color while other marbles have bands or veins of different colors. For example, it is common for white marble to contain dark bands.

Note that the word "marble" is used in the quarrying and decorative stone industry to refer to a large number of rocks, many of which are not actually marbles. For example, many decorative stones described as "black marbles" are actually dark limestones. Geologists only use the term marble to refer to a rock that formed through the metamorphism of carbonate-rich rocks.

Marble is found in many places, since limestone and dolostone are common rocks and are commonly metamorphosed. Since marble is often formed through regional metamorphism, it is commonly found near other metamorphic rocks, such as schists and gneisses. When marble forms through contact metamorphism, it can be found next to unmetamorphosed carbonate rocks.

In the United States, marble can be found in many states. Marble is quarried in several states, including Vermont, Colorado, Alabama, and Georgia. The largest quarry is the Danby Quarry in Vermont, which has operated for more than 100 years. This quarry produces several types of beautiful white decorative stone, which are used in houses and other buildings. Many kitchen countertops in the United States are made out of Danby Quarry marbles!

IDENTIFICATION: If you see a light-colored rock with large, sparkly crystals that are similar in size, then it could be a marble. Note that marble is commonly white but can be other colors, such as pink. Marble can look similar to limestone or dolostone, but it will have larger crystals that are interlocking and will be less porous. In addition, marble will not contain any fossils. Marble can also look like quartzite (a rock formed from the metamorphism of sandstone), but marble is much softer than quartzite and can be easily scratched with a knife. In addition, marble will fizz when a few drops of dilute acid (such as vinegar) are dropped onto the rock whereas a quartzite will not fizz. Sometimes, the individual crystals of marble are difficult to make out because they are all similar in color.

FUN FACT: Marble was used in many classical Greek and Roman sculptures and continues to be used in sculptures to the present day. For example, Michelangelo's David statue was carved from white marble.

SERPENTINITE

ROCK TYPE: Metamorphic

COLOR: Usually green

MINERALOGY: Mostly serpentine

TEXTURE: Variable grain size and variable foliation

DESCRIPTION AND OCCURRENCE: Serpentinite is formed through low-grade metamorphism of mafic (silica-poor) or ultramafic (very silica-poor) rock, usually basalt or peridotite. Serpentinite forms when the original mafic minerals (mostly olivine and pyroxene) react with water-rich metamorphic fluids and transform into secondary alteration minerals. The primary alteration minerals that form are green serpentine group minerals, such as lizardite and chrysotile. White carbonate minerals and white talc are also commonly formed. Other minerals, such as magnetite and brucite, also sometimes form. In addition, some of the original ultramafic minerals, such as olivine, pyroxene, and chromite, may survive in the altered rock. The process of formation of a serpentinite rock is known as serpentinization. Commonly, peridotites will be partially serpentinized. This means that portions of the rock or rock outcrop have been altered, depending on where metamorphic fluids flowed through the rock.

The texture of serpentinite is variable, depending on the degree of serpentinization and the types of secondary minerals that form. The grain size of serpentinite ranges from small to large. Some serpentinites display foliation (layering), while others are non-foliated. The secondary minerals commonly form as veins in the original peridotite rock. Serpentinite is usually green in color and often has streaks or veins of white. Sometimes, serpentinite can be yellow or brown in color.

On land serpentinites are generally found in ophiolites, which are sections of ocean crust and underlying mantle (which mostly consists of peridotite) that have been uplifted and exposed on land. You can also find serpentinites on the ocean floor, where faulting in the ocean crust exposes underlying peridotite rock. Geologists are particularly interested in studying the formation of serpentinite on the seafloor because some scientists believe that life on Earth could have originated billions of years ago at a hydrothermal vent (a location on the seafloor where hot water discharges) located in serpentinite rock. Hydrothermal vents associated with serpentinite rock on the seafloor produce special chemicals and chemical gradients that could have created the perfect conditions for the formation of the first microscopic life.

In the United States, two good places to find serpentinite are Vermont and California. In Vermont, slivers of serpentinite can be found in the Green Mountains, although they can be difficult to spot among all of the green vegetation. Serpentinite is a little easier to see in California, where it can be found in several locations, such as in the Golden Gate National Recreation Area in San Francisco. Serpentinite is even the official state rock of California.

Some types of serpentinite contain asbestos, which is not a mineral but rather a term used to describe minerals when they have an elongated, fibrous form. Sometimes, the serpentine group mineral chrysotile forms in an asbestos shape. Before the health impacts of asbestos use were known, asbestos was mined from serpentinites around the world for use in building materials and many other products. Historically, asbestos was mined from serpentinite at dozens of locations in the United States, for example at Belvidere Asbestos Mine in Vermont and at King City Asbestos Mine in California.

IDENTIFICATION: Serpentinite can be a little challenging to identify because it has a variable appearance. However, it is generally green in color and commonly contains veins of secondary minerals, which gives the rock a snakeskin like appearance. The serpentine minerals often give serpentinite a shiny or waxy appearance. Serpentinite commonly contains talc, which can make the rock feel slippery like soap. Although serpentinite can contain asbestos minerals, it rarely poses a health risk when viewed in an outcrop. However, geologists need to be careful when crushing the rock for analysis, since this can release potentially dangerous small asbestos fragments.

FUN FACT: The term serpentinite comes from the word "serpentine" (snake-like), since the color and texture of many serpentinite rocks looks similar to a snakeskin.

ECLOGITE

ROCK TYPE: Metamorphic

COLOR: Red, green, and white

MINERALOGY: Mostly garnet and pyroxene

TEXTURE: Medium- to coarse-grained, usually non-foliated

DESCRIPTION AND OCCURRENCE: Eclogite is a medium- to coarse-grained metamorphic rock that primarily consists of red garnet (almandine-pyrope variety) and green pyroxene (omphacite variety). Eclogite forms under high pressure and high temperature conditions from basalt and gabbro that are brought down into the Earth's mantle at a subduction zone, for example when denser oceanic crust is subducted underneath less dense continental crust. Eclogite often contains white quartz, so the rock is commonly red, green, and white in color. Eclogite can also contain small amounts of other minerals that form during the initial metamorphism or during mineralogical changes that occur as the rock is brought from deep in the Earth up to Earth's surface (this is called "retrograde metamorphism"). Examples of other minerals that can be found in eclogite are kyanite, rutile, and amphibole. Rarely, eclogites contain diamonds.

Texturally, eclogites are usually non-foliated. However, they can display foliation (layering), for example when they occur in shear zones (an area where rocks have been deformed due to movement along a geologic fault). Eclogite has the same chemical composition as basalt and gabbro, which are mafic rocks with a high amount of iron and magnesium. Eclogite is a very dense rock, with a density of about 3.5 to 3.7 g/cm^3.

Eclogite is a common rock in parts of Earth's mantle. However, it is an uncommon rock at Earth's surface. Eclogite is generally found where part of the Earth's mantle is brought to the surface, for example at an ophiolite (a section of oceanic crust and underlying mantle exposed on land through tectonic processes). Thus, eclogites are found in association with other ophiolite rocks, such as peridotites and serpentinites. Eclogites can also be found as xenoliths (literally "foreign rocks") in certain types of volcanic rock. For example, they can be found in kimberlite, which forms from magmas that originate deep in the Earth's mantle.

In the United States, the best place to see eclogites is in the California Coast Range, a mountain range stretching from northern California down to Santa Barbara. Eclogites can be seen in the Franciscan Complex rocks found in this mountain range, for example near San Francisco. Eclogites can also be found in areas of North Carolina and South Carolina in the Appalachian Mountains. Another good place to look for eclogite is a museum. Because eclogite is such a beautiful and interesting rock, many museums have eclogite specimens on display.

IDENTIFICATION: Eclogite is an easy rock to recognize. If you see a rock with coarse crystals of red garnet and green pyroxene, then it is most likely an eclogite. There may be some white minerals, such as quartz, present as well. Eclogite is a very dense rock, so it will feel heavy compared to most other rocks.

FUN FACT: Geologists commonly refer to eclogite as "Christmas rock" due to its having Christmas colors of red, green, and white.

RARE ROCKS

KIMBERLITE

ROCK TYPE: Igneous

COLOR: Generally blue-gray or green

MINERALOGY: Primarily olivine (often altered to serpentine), mica (phlogopite variety), pyroxene, and garnet

TEXTURE: Usually porphyritic

DESCRIPTION AND OCCURRENCE: Kimberlite is a rare igneous rock that is ultramafic in composition, which means that it is enriched in magnesium and iron and has a relatively low amount (compared to other rocks) of silica. Kimberlite is highly enriched in volatiles, such as carbon dioxide and water. Kimberlite magmas originate very deep in the Earth at depths greater than 150 kilometers. Kimberlite magmas form through very small amounts of melting, which makes them enriched in incompatible elements (such as potassium), which are elements that prefer to go into a melt compared to staying in the rock that is melted.

Kimberlite is a porphyritic rock and consists of phenocrysts (large crystals) set in a finer-grained matrix. The phenocrysts are most commonly olivine crystals, which have often been partly or completely altered to serpentine minerals. The phenocrysts can also be other minerals, such as mica (phlogopite variety), garnet, pyroxene, ilmenite, and spinel. The finer-grained matrix (or groundmass) of kimberlite consists largely of olivine, along with other minerals, such as mica (phlogopite variety), pyroxene, apatite, spinel, perovskite, and ilmenite. The matrix also commonly contains serpentine and carbonate alteration minerals. Very commonly, kimberlite contains xenoliths (literally "foreign rocks"), which can be mantle rocks, such as peridotite and eclogite, or crustal rocks, including the "country rocks" through which the kimberlite erupts. The xenoliths are rock pieces that were ripped up and included in the kimberlite magma as it ascended to Earth's surface. Kimberlite can also contain xenocrysts (literally "foreign crystals"). Rarely, kimberlite contains xenocrysts of diamonds, which are brought up from deep in the Earth's mantle.

No one has ever observed a kimberlite volcano erupt. Potentially, the most recent kimberlite eruption occurred approximately 12,000 years ago at Igwisi Hills in Tanzania. However, geologists are not certain if the Igwisi Hills rocks are true kimberlites. The next youngest kimberlite volcano erupted more than 30 million years ago in the Democratic Republic of the Congo. Many kimberlites erupted between about 70 and 150 million years ago, and some are more than 1 billion years old. Although no one has ever observed a kimberlite eruption, geologists believe that kimberlite eruptions occurred quickly and explosively, with the magmas traveling at great speeds up through Earth's crust.

Small volumes of kimberlite rock have been found on every continent, including Antarctica. Kimberlite rock is often associated with large crustal structures, such as sutures between continental blocks, which may have provided pathways for the magma to rise from great depths. Kimberlite is often found in vertical structures called pipes. Approximately 7,000 kimberlite pipes are known. However, the majority of them do not contain diamonds. Less than 100 kimberlite pipes have been mined for diamonds, and only a handful are "Tier 1" mines that produce large quantities of diamonds. Examples of large kimberlite diamond mines are the Cullinan and Venetia mines in South Africa, the Jwaneng and Orapa mines in Botswana, the Catoca Mine in Angola, and the Aykhal and Mir mines in Russia.

In the United States, small kimberlite pipes can be found in several states, including Kansas, Colorado, and Wyoming. Two kimberlite pipes in Colorado were briefly mined for diamonds from 1996 to 2001. The only other place where diamond has been mined in the United States is in Arkansas, where the diamonds are found in lamproite rather than kimberlite. Lamproite is also a rare ultramafic igneous rock but has a slightly different chemical and mineralogical composition than kimberlite.

IDENTIFICATION: Kimberlite is one of the more difficult rocks to identify, because it is often highly altered and contains xenoliths, which can be both mantle rocks and pieces of Earth's crust. Generally, kimberlites will have a porphyritic texture with large mineral grains and xenoliths in a finer-grained matrix. Kimberlite often has a blue-gray or green color but can be yellow or yellowish brown when highly weathered. In order to identify a kimberlite, geologists usually examine thin slices of the rock under a microscope.

FUN FACT: Kimberlite is named after the town of Kimberley, South Africa, where diamonds have been mined since the late 1800s.

METEORITE

ROCK TYPE: Variable

COLOR: Usually gray or brown with a black surface

MINERALOGY: Variable

TEXTURE: Variable

DESCRIPTION AND OCCURRENCE: A meteorite is a rock that originates in outer space and falls to Earth. Meteorites are heavy, dark-colored rocks. They commonly have a black fusion crust that forms when the meteorite is heated as it falls through Earth's atmosphere. They can also have regmaglypts, which are rounded depressions that form on the meteorite's surface when it is partially melted during its descent to Earth. Due to the presence of iron-rich metal, meteorites are magnetic. Meteorites can be any size, from micrometeorites that are less than a millimeter in size to large meteorites that are several meters in size.

Meteorites are variable in mineralogy and chemical composition. Planetary scientists have many terms to describe meteorites, based on their chemistries, textures, and origins. A simple but useful classification for beginners is to group them into three basic categories: iron meteorites, stony-iron meteorites, and stony meteorites.

Iron meteorites consist mostly of nickel-iron alloys (metals) and are believed to represent planetary cores, since iron and other heavy elements sink to the centers of planetary bodies. The core of our own Earth is believed to have a composition similar to an iron meteorite. When cut to expose a fresh (not weathered or obscured by a fusion crust) surface, an iron meteorite looks like gray or silvery metal. Many iron meteorites display an interesting property: when etched with acid, a criss-crossing pattern called a "Widmanstätten Pattern" appears. This pattern is found in certain meteorites and represents the intergrowth of two nickel-iron alloy minerals called kamacite and taenite.

Stony-iron meteorites are composed of nickel-iron alloys and silicate minerals in about equal proportions. The silicate minerals are generally olivine, plagioclase, or pyroxene. Scientists are not completely sure how stony-iron meteorites formed, but some may have originated from the core-mantle boundary of a planetary body. Others likely formed when two asteroids collided.

Stony meteorites consist mostly of silicate minerals. The mineralogy of stony meteorites varies, but they commonly contain plagioclase, pyroxene, and olivine. They may contain small amounts of nickel-iron alloys. There are two sub-types of stony meteorites: chondrites and achondrites. Chondrites contain chondrules, which are small spheres that formed from molten material during the early formation of proto-planetary bodies. Chondrites are more than 4.5 billion years old and represent the oldest material found in our solar system. Analysis of chondrites gives planetary scientists important clues about the nature of the initial material that accreted and differentiated (separated into the core, mantle, and crust) to form the Earth and the other rocky planets. Another type of stony meteorite is an achondrite, which does not contain chondrules. Achondrites are believed to be igneous rocks that formed from melting that occurred in planetary bodies.

Most meteorites originate from the asteroid belt located in between Mars and Jupiter. However, a few meteorites are known to have come from the Moon and Mars.

Meteorites can fall anywhere on Earth. So, in theory you can find a meteorite anywhere! However, meteorites resemble Earth rocks. So, it can be difficult to spot a meteorite among the more abundant Earth rocks. Two places where it is easier to find meteorites are deserts (both hot deserts and cold deserts, for example those located in Antarctica) and areas with ice and snow. This is because it is easier to spot meteorites when there is no vegetation and few other rocks. In addition, in deserts the dry climate helps prevent meteorites from weathering. In the polar regions, meteorites can become encased in ice, which also helps with preservation. Movement of the ice can also concentrate meteorites in certain parts of an ice sheet.

Meteorites are very rare, so you are unlikely to find one. The best place to see meteorites is in a museum collection, for example at the Smithsonian Institution in Washington, D.C., or at the American Museum of Natural History in New York City. You could also visit a meteor crater, a depression resulting from a meteorite impact. There is an amazing example of one of these depressions at Meteor Crater (also called Barringer Crater) near Flagstaff, Arizona. This enormous meteor crater is about 4,000 feet in diameter and over 500 feet deep! It was formed by an impact that occurred approximately 50,000 years ago.

IDENTIFICATION: If you find a dark-colored, heavy rock that has a bumpy shape, then it could be a meteorite. Meteorites are magnetic, so they will attract a magnet. However, use of strong magnets can erase valuable scientific information from meteorites, so try not to use this identification method if you can avoid it. You can also identify a meteorite by looking for a dark fusion crust and curved depressions in the rock that could be regmaglypts. However, bear in mind that meteorites are very rare and there are Earth rocks and man-made materials, such as slag, that look similar to meteorites. So, don't get your hopes up! Meteorites will not have any vesicles (holes) in them, so if you see bubble-like holes, then the rock is probably a volcanic rock or a slag. The most definitive way to identify a meteorite is to analyze its chemical composition, which will need to be done at a university or museum.

FUN FACT: The largest meteorite ever found is the Hoba Meteorite, which weighs more than 60 tons! This meteorite was found near the town of Grootfontein in Namibia. You can visit the meteorite as part of a tourist attraction.

TEKTITE

ROCK TYPE: Metamorphic

COLOR: Black, brown, green, or gray

MINERALOGY: Silica glass (mineraloid)

TEXTURE: Glassy

DESCRIPTION AND OCCURRENCE: Tektites are glassy rocks that are formed by the melting of Earth rocks by the impact of large meteorites. Tektites are composed of fairly homogeneous silica-rich glass. Often, they contain a mineraloid called lechatelierite, which is an amorphous type of silica that forms when quartz is melted at high pressure. Sometimes, tektites contain inclusions of partly melted (and sometimes shocked) grains that are remnants from the original rocks that melted. For example, tektites can contain inclusions of quartz, apatite, and zircon. Tektites have a chemical composition similar to the rocks which melted to form them, except that they have a lower content of volatile molecules, such as water and carbon dioxide. This is because the volatiles are released during the melting of the rock.

Tektites can be black, brown, green, or gray in color and can be translucent or opaque. Tektites can have a variety of morphologies, including sphere, teardrop, button, and dumbbell shapes. Tektites generally range in size from a few millimeters to a few centimeters. Tektites that are less than 1 mm in size are called "microtektites."

Tektites have been found all over the world. They are found in strewn fields created by meteorite impacts. Notable tektite strewn fields are located in Australia, the Philippines, central Europe, Ivory Coast, and the United States. Tektites are commonly named by the areas where they are found. For example, tektites found in Australia are called "australites" and tektites found in the Philippines are called "philippinites." In the United States, there is a strewn field of tektites near the town of Bedias in Texas. These tektites are called "bediasites." There is also a strewn field in the state of Georgia, where the tektites are called "georgiaites." The tektites in Texas and Georgia are both believed to have been created by a meteorite that landed in the Chesapeake Bay area about 35 million years ago.

IDENTIFICATION: Tektites are rare rocks, so you are unlikely to find one. If you find a glassy rock, it is more likely a piece of volcanic obsidian or possibly man-made glass or slag. If you do find a rock that you suspect is a tektite, take it to a university or museum, where a geologist will be able to identify it. Tektites can be identified through chemical analysis. They will have very low water content compared to obsidian. Tektites also usually have a lower silica content than obsidian. The presence of lechatelierite and inclusions of shocked or partially melted mineral grains can also help confirm that a rock is a tektite.

FUN FACT: The word "tektite" comes from the Greek word "tektos," which means "molten."

FULGURITE

ROCK TYPE: Metamorphic

COLOR: Variable

MINERALOGY: Mostly glass (mineraloid)

TEXTURE: Variable

DESCRIPTION AND OCCURRENCE: Fulgurite is a rare rock that forms when lightning strikes sediment, soil, or rock, melting some of the preexisting material and turning it into a glass. Fulgurites often contain some unmelted rock fragments that become incorporated into the glass.

One type of fulgurite forms when lightning strikes sand. Since sand is mostly composed of silica-rich quartz, this type of fulgurite generally contains lechatelierite, a type of silica-rich glass that can also form as a result of a meteorite impact or a nuclear bomb explosion. Sand fulgurites usually have a glassy interior and a bumpy exterior from grains of sand that become fused to the glass. They commonly have a long, tube-like shape, although they can also form as flat clumps of material. Sand fulgurites are variable in color but are often white, brown, or black. Most sand fulgurites are only a few centimeters in size. However, some of the biggest ones are four or five meters long!

Another type of fulgurite forms when hard rock is struck by lightning, generally at the top of a mountain. These fulgurites form a smooth crust of glass on top of the rock. Fulgurites can also form when other types of sediment (other than sand) or soil are struck by lightning. Note that fulgurites do not have a fixed mineralogical or chemical composition, since their composition depends on the material that was melted by the lightning.

IDENTIFICATION: Glassy material found in an open area where lightning could have struck could be a fulgurite. You are most likely to find fulgurite on a beach or in a sandy desert environment, where it may look like a bumpy tree root. Sand fulgurites tend to be fragile, so be careful when picking one up or it may crumble in your hand. You may also find fulgurite at the top of a mountain, where exposed rock may have a glassy crust. Fulgurites are hard to find, so your best bet is probably to look for one in a museum!

FUN FACT: The word "fulgurite" comes from "fulgur," the Latin word for lightning.

COPROLITE

ROCK TYPE: Sedimentary (fossil)

COLOR: Variable—often brown, black, or bhite

MINERALOGY: Variable—generally phosphate, silicate, and/or carbonate minerals

TEXTURE: Variable—surface can be rough or smooth

DESCRIPTION AND OCCURRENCE: Coprolites are fossilized feces (poo), generally from animals that lived millions of years ago, including dinosaurs, sharks, and marine reptiles. Coprolites are rare because feces usually decay before they can be fossilized. However, if feces are quickly covered by sediment in an oxygen-poor environment, coprolite fossils may form. Coprolites are generally found in sedimentary rocks, such as limestone, chalk, and mudstone. They are almost always found along with other animal fossils, such as bones and teeth. Coprolites are more commonly preserved in marine environments than in terrestrial environments. In addition, feces from a carnivore are more likely to turn into a coprolite than feces from an herbivore. This is because carnivore feces contain calcium- and phosphate-rich bones and teeth that decay more slowly than plant matter.

Coprolites are rare, so you are unlikely to find one yourself. However, many rock and fossil shops offer coprolites for sale. You can also see them in many museums. The intrepid collector may have luck finding coprolites in Cretaceous-age (66- to 145-million-year-old) rocks in the central United States. For example, shark, fish, and marine reptile coprolites can be found in the Niobrara Chalk in Kansas, and dinosaur coprolites can be found in the Hell Creek Formation in South Dakota.

IDENTIFICATION: Coprolites usually look like feces, so they are fairly easy to identify. However, you need to be careful because other types of fossils and rocks can also look like feces. For example, fossilized tree roots and some types of mineral concretions can also look like feces. Generally, coprolites are found near other animal fossils, such as bones and teeth. In addition, they often contain inclusions of food matter, such as bones, teeth, and scales. Coprolites range from very small (about a millimeter in size) to very large (tens of centimeters in size), depending on the size of the animal that produced the feces.

FUN FACT: Mary Anning, a famous fossil hunter and paleontologist, provided key observations that led to the discovery of coprolites. In the 1820s she found and studied many ichthyosaur (marine reptile) fossils along the southwest coast of England. Near some of these skeletons she found stones that, when broken open, contained fossilized fish bones and scales. Her observations led geologist William Buckland to propose that these stones are fossilized feces and to coin the term "coprolite" (meaning "dung stone" in Greek) to describe them.

COMMON MINERALS

QUARTZ

FORMULA: SiO_2

MINERAL CLASS: Silicate

CRYSTAL SYSTEM: Trigonal

COLOR: Variable—often colorless, white, purple, pink, or black

TRANSPARENCY: Variable

STREAK: White

LUSTER: Usually vitreous

MOHS SCALE HARDNESS: 7

SPECIFIC GRAVITY: 2.6–2.7

DESCRIPTION AND OCCURRENCE: Quartz is a very common mineral composed of silica and oxygen. Quartz occurs in a wide variety of igneous, sedimentary, and metamorphic rocks. Quartz is particularly common in granite, pegmatite, sandstone, and quartzite (metamorphosed sandstone). Quartz is a highly durable mineral that is resistant to weathering, so desert, river, and beach sand is commonly composed largely of quartz.

Quartz crystals range in size from microscopic to very large. Some pegmatites contain quartz crystals that are several meters in length. When quartz grows into open cavities, it forms as hexagonal crystals that have a six-sided pyramid at one end. Quartz that forms in cavities sometimes creates a geode, which can be broken open to reveal the sparkling crystals inside. Quartz is also commonly found in massive form, with no visible crystals.

Quartz comes in a wide variety of colors and is known by many different names depending on the color. Clear quartz is often referred to as rock crystal. Examples of colored varieties of quartz are milky quartz (white), smoky quartz (gray, brown, or black), amethyst (purple), rose quartz (pink), and citrine (yellow). Quartz sometimes contains inclusions of other minerals, most commonly needle-shaped crystals of rutile (rutilated quartz). Tiger's Eye is a beautiful golden gemstone that is formed when quartz replaces the mineral crocidolite during metamorphism.

Chalcedony, which can be seen in banded agate rocks (including onyx) and red jasper, is closely related to quartz. Chalcedony consists of very small (cryptocrystalline) grains of quartz and moganite, another mineral with a different crystal structure but the same chemical composition as quartz. Although chalcedony technically consists of two minerals, it is often described as a single mineral, for example in reference books. It is sometimes classified as a type of quartz. The rocks flint and chert, which have been used to make tools throughout human history, are also composed of cryptocrystalline quartz.

Quartz can be found almost everywhere. Milky quartz is most common and can be found as small stones or rounded pebbles that have weathered out of rocks. Quartz crystals are easy to purchase in rock shops or museum gift shops. In the United States, there are locations where you can dig for your own impressive quartz crystals. One of the best places to dig for quartz crystals is the Ouachita Mountains in Arkansas, where there are several quartz crystal mines that are open to the public for a fee. Another good place to find beautiful quartz crystals is Herkimer, New York, where you can find clear quartz crystals that, unusually, have six-sided pyramids on both ends (rather than just on one end). These quartz crystals superficially resemble diamonds and are known as "Herkimer diamonds."

IDENTIFICATION: When crystals are present, quartz can be identified by its characteristic crystal shape: a hexagonal crystal topped with a six-sided pyramid. Quartz can also be identified by its vitreous luster and the presence of conchoidal (curved) fracture surfaces. Massive clear or white quartz can be confused with calcite, but calcite is much softer (hardness of 3). In addition, calcite will fizz when a few drops of dilute acid (such as vinegar) are dropped onto the rock whereas quartz will not fizz.

FUN FACT: An enormous cluster of quartz crystals called the "Berns Quartz" can be viewed at the Smithsonian Institution in Washington, D.C. This quartz cluster was discovered in a mine in Arkansas in 2016 and weighs over 8,000 pounds!

FELDSPAR

FORMULA: $X(Al,Si)_4O_8$

MINERAL CLASS: Silicate

CRYSTAL SYSTEM: Monoclinic or triclinic

COLOR: Variable—often white, gray, or pink

TRANSPARENCY: Variable

STREAK: White

LUSTER: Usually vitreous, sometimes pearly

MOHS SCALE HARDNESS: 6.0–6.5

SPECIFIC GRAVITY: 2.5–2.6

DESCRIPTION AND OCCURRENCE: Feldspar is not a single mineral but rather a group of closely related silicate minerals. The general formula for feldspar is $X(Al,Si)_4O_8$, where X is generally potassium (K), sodium (Na), or calcium (Ca). More rarely, X can be barium (Ba). The feldspars are described in terms of three chemical composition end members: orthoclase ($KAlSi_3O_8$, also known as "K-feldspar" or "K-spar"), albite ($NaAlSi_3O_8$), and anorthite ($CaAl_2Si_2O_8$). The chemical composition of feldspar can vary between orthoclase and albite (alkali feldspars) and also between albite and anorthite (plagioclase feldspars). The term "solid solution" is used to describe the chemical variation, where different proportions of chemical elements can fill the X position of the mineral formula. Feldspar minerals also vary in their crystal structure, which can be either monoclinic or triclinic. There are many different feldspar minerals. Examples of alkali feldspar minerals are sanidine and microcline. Examples of plagioclase feldspar minerals are oligoclase, andesine, labradorite, and bytownite.

Despite their chemical and structural variability, minerals in the feldspar group have some common characteristics. They have a hardness of 6.0–6.5 and a specific gravity of about 2.5–2.6. In addition, they have two cleavage planes (surfaces along which a mineral tends to break) that are perpendicular to each other. Thus, feldspar crystals tend to break into blocks with 90 degree angles. Feldspar minerals usually have a vitreous luster, although they can also have a pearly luster. Plagioclase feldspars often have fine parallel lines running across their crystal surfaces or cleavage planes. These lines result from crystal twinning, which occurs when a crystal continues growing on itself but has switched its orientation causing the crystal to grow in a different direction.

The color of feldspar is variable. Feldspar is commonly white or gray. Potassium-rich feldspar is often pink or red in color. The alkali feldspar microcline is sometimes a green or blue-green color. This type of microcline is referred to as amazonite. The plagioclase feldspar labradorite sometimes displays beautiful iridescence with blue, green, yellow, and orange colors.

Taken as a group, feldspar is the most common mineral found in the Earth's crust. Feldspar minerals are found in a wide variety of igneous, sedimentary, and metamorphic rocks. In general, the more calcium-rich feldspars are more common in mafic (silica-poor) rocks compared to felsic (silica-rich) rocks. In sediments and sedimentary rocks, feldspar minerals are easily altered to clay minerals.

Feldspar can be found almost everywhere in the United States. Many common rocks that you see, such as granite, contain feldspar. If you want to find an impressive feldspar crystal for your mineral collection, large feldspar crystals can be found in pegmatites. You can find pegmatites with feldspars in nearly all of the major mountain ranges in the United States, with some impressive examples occurring in Maine, New Hampshire, Virginia, South Dakota, Colorado, New Mexico, and California. Green amazonite is a particularly beautiful feldspar to add to your collection, but it is rare. In the United States, amazonite can be found in the state of Colorado, for example near Pikes Peak.

IDENTIFICATION: Feldspar minerals can be identified by their hardness and tendency to break into blocks that have 90 degree angles. Identification of specific feldspar minerals can be challenging without a microscope or chemical analysis. However, potassium-rich feldspar is often pink or red in color. Amazonite can be recognized by its distinctive green or blue-green color, and labradorite can sometimes be recognized by its iridescence. Plagioclase feldspars often have thin parallel lines on mineral surfaces. These lines are not present on alkali feldspars.

FUN FACT: Feldspar is a very abundant mineral on the moon. The lunar highlands, which are the lighter-colored areas of the moon, are composed of a rock called anorthosite, which is made up of 90% or more plagioclase feldspar.

PYROXENE

FORMULA: $XY(Si, Al)_2O_6$

MINERAL CLASS: Silicate

CRYSTAL SYSTEM: Monoclinic or orthorhombic

COLOR: Variable—often dark green or black

TRANSPARENCY: Translucent to opaque

STREAK: White, gray, or pale green

LUSTER: Usually vitreous

MOHS SCALE HARDNESS: 5–7

SPECIFIC GRAVITY: 3–4

DESCRIPTION AND OCCURRENCE: Pyroxene is not a single mineral but rather a group of closely related silicate minerals. The general chemical formula for pyroxene is $XY(Si, Al)_2O_6$, where X and Y are generally the elements sodium (Na), calcium (Ca), iron (Fe), magnesium (Mg), manganese (Mn), lithium (Li), zinc (Zn), aluminum (Al), chromium (Cr), or titanium (Ti).

There are two main types of pyroxenes: clinopyroxenes, which crystallize in the monoclinic crystal system, and orthopyroxenes, which crystallize in the orthorhombic mineral system. There are many different pyroxene minerals. Examples of clinopyroxene minerals are augite and diopside. Examples of orthopyroxene minerals are enstatite and ferrosilite.

Despite their chemical and structural variability, minerals in the pyroxene group have some common characteristics. They usually have a vitreous luster and a small range of hardness and specific gravity. In addition, they have two cleavage planes (surfaces along which a mineral tends to break) that are perpendicular to each other. Thus, pyroxene crystals tend to break into blocks with 90 degree angles. Pyroxene minerals are most commonly dark green or black in color. However, they can be other colors as well, including light apple green, white, and brown. Spodumene, a lithium-rich pyroxene, can be a beautiful lilac purple color.

Pyroxene minerals are found in a wide variety of igneous and metamorphic rocks. They tend not to survive in sediments and sedimentary rocks because they are easily weathered. Pyroxene minerals are particularly abundant in mafic (silica-poor) igneous rocks, such as basalt and peridotite. A rock called pyroxenite consists mostly of pyroxene minerals. Pyroxenite is an ultramafic (very silica-poor) rock that is common in Earth's mantle but is uncommon in Earth's crust. Near Earth's surface, large amounts of pyroxenite can be found in layered igneous intrusions, which are rare. The pyroxenite found in layered igneous intrusions can contain high amounts of rare elements—such as platinum, chromium, and nickel—and is often mined. The biggest layered igneous intrusion is the Bushveld Igneous Complex in South Africa. In the United States, pyroxenite can be found in the Stillwater Igneous Complex, a layered igneous intrusion located in the state of Montana.

IDENTIFICATION: Pyroxene minerals can be identified by their vitreous luster and dark green or black color, although pyroxene can sometimes be other colors, such as bright apple green or white. Pyroxene minerals can look similar to amphibole minerals. However, a good way to tell pyroxene and amphibole apart is the difference in the angles of their intersecting cleavage planes (surfaces along which a mineral tends to break). Pyroxene minerals have two cleavage planes that intersect at roughly 90 degree angles, which means that they tend to break into blocks that have 90 degree angles. Amphibole minerals, on the other hand, have cleavage planes that intersect at a different angle, and they tend to break into blocks that have roughly 60 degree and 120 degree angles. Feldspar minerals also break into blocks with 90 degree angles, but they are usually lighter in color than pyroxene minerals.

FUN FACT: The semi-precious green gemstone jade is one of two minerals: jadeite, a sodium-rich clinopyroxene mineral, or nephrite, an amphibole mineral. The two minerals look very similar and are difficult to distinguish with the naked eye.

AMPHIBOLE

FORMULA: $W_{0-1}X_2Y_5Z_8O_{22}(OH,F)_2$

MINERAL CLASS: Silicate

CRYSTAL SYSTEM: Usually monoclinic

COLOR: Variable—often brown, white, black, or dark green

TRANSPARENCY: Variable

STREAK: Variable—often white

LUSTER: Variable—often vitreous

MOHS SCALE HARDNESS: 5–6

SPECIFIC GRAVITY: 2.8–3.6

DESCRIPTION AND OCCURRENCE: Amphibole is not a single mineral but rather a large group of related hydrous (water-bearing) silicate minerals. The general formula for amphibole is $W_{0-1}X_2Y_5Z_8O_{22}(OH,F)_2$. The chemical composition of amphibole varies greatly, depending on which elements are present in the W, X, Y, and Z positions of the mineral formula. There are more than 100 different amphibole minerals! Examples of common amphibole minerals are anthophyllite ($(Mg,Fe)_7Si_8O_{22}(OH)_2$), cummingtonite ($Fe_2Mg_5Si_8O_{22}(OH)_2$), tremolite

($Ca_2Mg_5Si_8O_{22}(OH)_2$), actinolite ($Ca_2(Mg,Fe)_5Si_8O_{22}(OH)_2$), hornblende ($Ca_2(Mg,Fe,Al)_5(Al,Si)_8O_{22}(OH)_2$), and glaucophane ($Na_2Mg_3Al_2Si_8O_{22}(OH)_2$).

Despite their wide variation in chemical composition, minerals in the amphibole group do have some common characteristics. One common characteristic of amphibole minerals is that they have two cleavage planes (surfaces along which a mineral tends to break) that intersect to form approximately 60 degree and 120 degree angles, which means they tend to break into blocks with these angles. Another common characteristic is that amphibole minerals all have a hardness of about 5–6. Amphibole minerals also usually have a vitreous luster.

However, different minerals in the amphibole group can look very different. The specific gravity of amphibole minerals varies over a range, with iron-bearing amphiboles generally being heavier. The color of amphibole minerals is also variable, although they are most commonly white, brown, black, or dark green in color. Hornblende, a common amphibole mineral found in many igneous rocks, is often black in color. Tremolite tends to be a lighter color, such as white. Actinolite is usually light to dark green. Glaucophane is commonly a grayish blue color. Some types of amphibole, such as hornblende, have short prismatic crystals. Other types of amphibole, such as tremolite and actinolite, often form acicular (needle-like) or fibrous crystals.

When amphibole minerals have elongated, fibrous crystals, they can be classified as asbestos. Of the six minerals that can form asbestos-shaped crystals, five are amphibole minerals: actinolite, amosite, anthophyllite, crocidolite, and tremolite. The sixth asbestos mineral is chrysotile, which is a serpentine group mineral.

Amphibole minerals are found in many igneous and metamorphic rocks. The amphibole mineral hornblende is commonly found in igneous rocks, such as granite, dacite, and basalt. A wonderful example of this is found at Black Butte in the Shasta-Trinity National Forest of California, where the dacite lava has beautiful slender crystals of hornblende. Amphibolite is a type of metamorphic rock that consists largely of amphibole minerals. Amphibolites are produced through the metamorphism of mafic (silica-poor) igneous rocks, such as basalt. A beautiful example of amphibolite occurs in the Gore Mountain region of New York State, where the amphibolite contains large garnet crystals that are mined for use in abrasives. Amphibole minerals are not generally found in sediments and sedimentary rocks because they are easily weathered.

IDENTIFICATION: Identifying amphibole is a little tricky, due to its variable chemical composition and appearance. A dark-colored amphibole mineral called hornblende is often found in igneous rocks and could be mistaken for pyroxene. However, amphibole minerals tend to break into blocks that have approximately 60 degree and 120 degree angles while pyroxene minerals tend to break into blocks that have 90 degree angles. Some amphibole minerals can be identified by their acicular (needle-like) or fibrous minerals. Be careful with amphibole minerals that have fibrous crystals. They could be a type of asbestos.

FUN FACT: The word "amphibole" comes from the Greek word "amphibolos," which means ambiguous. The mineral group was named by French mineralogist René-Just Haüy in reference to the variety of chemical compositions and physical characteristics found in the amphibole group of minerals.

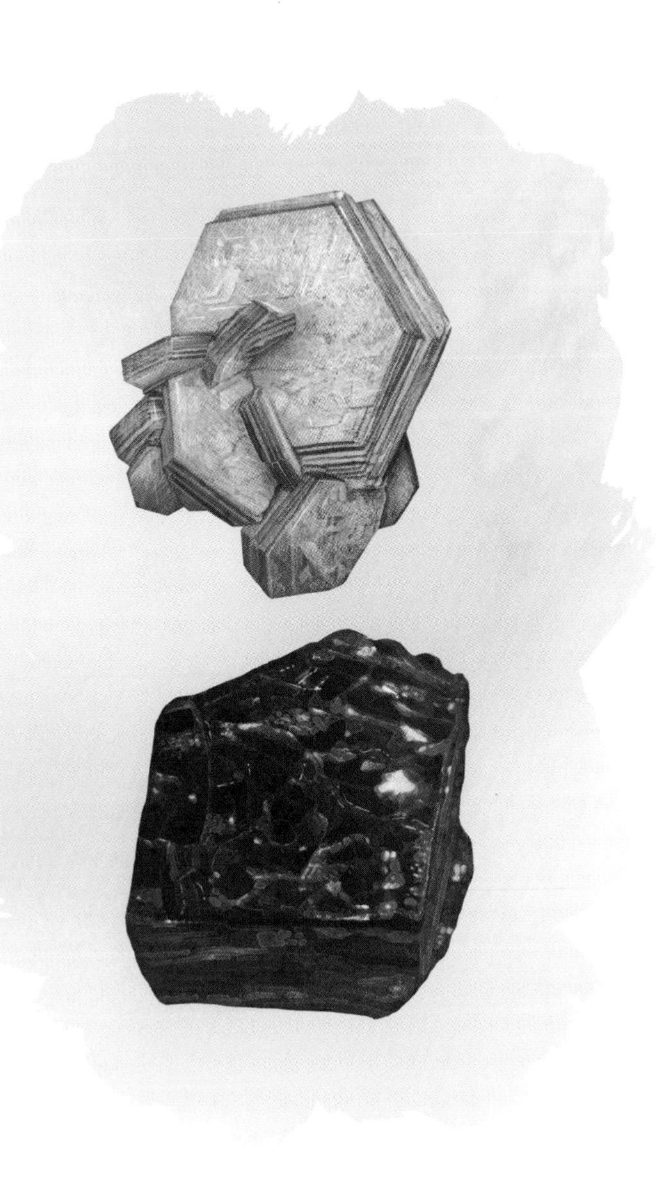

MICA

FORMULA: $XY_{2-3}Z_4O_{10}(OH,F)_2$

MINERAL CLASS: Silicate

CRYSTAL SYSTEM: Monoclinic

COLOR: Variable—commonly silver or black

TRANSPARENCY: Transparent to translucent

STREAK: White

LUSTER: Pearly or vitreous

MOHS SCALE HARDNESS: 2.5–4.0

SPECIFIC GRAVITY: 2.8–3.3

DESCRIPTION AND OCCURRENCE: Mica is a not a single mineral but rather a group of closely related silicate minerals. The general formula for mica is $XY_{2-3}Z_4O_{10}(OH,F)_2$, where X is commonly potassium (K), sodium (Na), or calcium (Ca), Y is commonly aluminum (Al), magnesium (Mg), or iron (Fe), and Z is commonly silicon (Si), aluminum (Al), or iron (Fe). Mica minerals can also contain other elements, such as lithium (Li), titanium (Ti), and chromium (Cr). Common mica minerals are muscovite, phlogopite, and biotite.

Despite their chemical variability, mica minerals share many common characteristics. Most notably, they easily break along cleavage planes (surfaces along which a mineral tends to break) into flat sheets. They also have a pearly or vitreous luster. When mica forms as crystals, they are usually hexagonal in shape. Stacks of large mica crystals are often called "books" because the thin layers of minerals resemble the pages of a book.

Different mica minerals can often be identified by their characteristic colors. Muscovite is a light-colored mica and is generally silver, gray, or white in color. Phlogopite and biotite are dark-colored micas. Phlogopite is usually brown in color while biotite is usually black in color. Mica minerals can also be many other colors. For example, a lithium-rich mica mineral called lepidolite is often a beautiful pale lilac color. As another example, a chromium-rich mica mineral called fuchsite is bright green in color.

Mica minerals are very common and can be found in many different types of igneous, sedimentary, and metamorphic rocks. In the United States, mica minerals can be found in many places. You are most likely to see a mica crystal in a coarse-grained igneous rock, such as granite, or in a mica-rich schist. If you want to find a large mica crystal to add to your collection, you can look for one in pegmatite rocks, which can be found in all the New England states. Lilac-colored lepidolite mica is found in pegmatites in the Black Hills of South Dakota, as well as at Harding Pegmatite Mine in New Mexico.

IDENTIFICATION: Mica minerals can be recognized by their pearly or vitreous luster and the fact that they can easily be split into thin sheets. You should be able to easily pull mica apart into sheets with your bare hands. You may see "books" or stacks of mica crystals. These crystals will often be hexagonal in shape. Muscovite, a common mica mineral, is usually a light color, such as silver. Phlogopite and biotite, two more common mica minerals, are usually brown or black in color.

FUN FACT: Historically, large, thin sheets of muscovite, a common mica mineral, were used in windows instead of glass. In fact, the word "muscovite" comes from "Muscovy glass," which was used to refer to large crystals of muscovite that were mined for this purpose near Moscow, Russia. Muscovy glass has even been found at the Jamestown Colony of present day Virginia, where it is thought to have been used in a lantern.

GARNET

FORMULA: $X_3Y_2Si_3O_{12}$

MINERAL CLASS: Silicate

CRYSTAL SYSTEM: Cubic

COLOR: Variable—often red

TRANSPARENCY: Variable

STREAK: White

LUSTER: Vitreous to resinous

MOHS SCALE HARDNESS: 6.5–7.5

SPECIFIC GRAVITY: 3.5–4.3

DESCRIPTION AND OCCURRENCE: Garnet is not a single mineral but rather a group of closely related silicate minerals. The general formula for garnet is $X_2Y_3Si_3O_{12}$, where X is commonly calcium (Ca), magnesium (Mg), manganese (Mn), or iron (Fe), and Y is commonly aluminum (Al), chromium (Cr), or iron (Fe). There are many different garnet minerals. However, garnets are often described in terms of chemical end members, which fall into two subgroups. The first is the pyrope ($Mg_3Al_2Si_3O_{12}$)–almandine ($Fe_3Al_2Si_3O_{12}$)–spessartine ($Mn_3Al_2Si_3O_{12}$) subgroup, and the second is the grossular ($Ca_3Al_2Si_3O_{12}$)–uvarovite ($Ca_3Cr_2Si_3O_{12}$)–andradite ($Ca_3Fe_2Si_3O_{12}$) subgroup. The garnet minerals in the first subgroup are called the "pyralspite garnets" and have aluminum in the Y position, while the garnet minerals in the second subgroup are called the "ugrandite garnets" and have calcium in the X position.

Despite their chemical variability, garnets share some common characteristics. They are similar in hardness and specific gravity, although manganese-rich and iron-rich garnets tend to be heavier. They have a vitreous to resinous luster. Garnet often forms as crystals, which commonly are in the shape of a rhombic dodecahedron, which is a twelve-sided shape with diamond-shaped (rhombic) faces.

Garnet is variable in color. Some of the more common garnets, such as pyrope and almandine, tend to be red, purple, or black in color. Spessartine garnet is often an orangish red color. Grossular garnet can be various colors, such as white, yellow, brown, and green. Andradite garnet can also be various colors, such as yellow, brown, and black. Uvarovite is usually a beautiful emerald green color. A variety of different garnet minerals of various colors are used as gemstones.

Garnet is commonly found in metamorphic rocks, such as schist and gneiss. Garnet can also be found in some types of igneous rocks, such as pegmatite, peridotite, and kimberlite. Certain garnet minerals tend to be found in certain rocks. For example, almandine garnet is commonly found in schist and gneiss while pyrope garnet is commonly found in peridotite and kimberlite. In addition, garnet can be found in sedimentary rocks and is often found in beach and river sand, where water movement and wave action often concentrate it along with other heavy minerals, such as magnetite.

In the United States, garnet is most easily found in metamorphic rocks. One spectacular location where you can find garnet is at Gore Mountain in the Adirondack Mountains of New York, where large garnets can be found in a metamorphic rock called amphibolite. The garnets found at Gore Mountain are unusually large, often 10–20 centimeters in diameter! The garnets at Gore Mountain were mined for use in abrasives for over 100 years, starting in the late 1800s. Today, garnets are still mined at nearby Ruby Mountain.

IDENTIFICATION: The easiest way to identify garnet is by its crystal shape, which is usually a rhombic dodecahedron (a twelve-sided shape with diamond-shaped faces). In addition, garnet can be recognized by its hardness and luster. Garnet comes in many different colors, but the most common color is red. Telling apart different minerals in the garnet group can be challenging and sometimes requires a chemical analysis. The type of rock in which the garnet is found can provide a clue as to which garnet is likely present. For example, igneous rocks will often contain pyrope garnet.

FUN FACT: In 1885 a giant garnet crystal the size of a bowling ball was found underneath New York City during excavations for a sewer system. The garnet was dubbed the "Subway Garnet" (because "Sewer Garnet" seemed too impolite) and is also called the "Kunz Garnet," after one of the first owners of the garnet. Today, the garnet is part of the mineral collection at The American Museum of Natural History in New York City.

OLIVINE

FORMULA: $(Mg, Fe)_2SiO_4$

MINERAL CLASS: Silicate

CRYSTAL SYSTEM: Orthorhombic

COLOR: Green or yellow-green

TRANSPARENCY: Transparent to translucent

STREAK: White

LUSTER: Vitreous

MOHS SCALE HARDNESS: 6.5–7

SPECIFIC GRAVITY: 3.2–4.4

DESCRIPTION AND OCCURRENCE: Olivine is a silicate mineral that is commonly found in igneous mafic and ultramafic rocks, which have a high amount of magnesium and iron and a low amount (relative to other rocks) of silica. Olivine can also be found in some metamorphic rocks that are produced from the metamorphism of magnesium-rich sediments and sedimentary rocks.

The proportion of magnesium and iron in olivine is variable. Olivine has two end member compositions: the magnesium-rich end member is called forsterite (formula: Mg_2SiO_4), and the iron-rich end member is called fayalite (formula: Fe_2SiO_4). Olivine crystals generally contain both magnesium and iron. However, olivine crystals that are rich in magnesium are described as "forsteritic olivine" while olivine crystals that are rich in iron are described as "fayalitic olivine." The specific gravity of olivine varies, with fayalite having a higher specific gravity because iron is heavier than magnesium.

Fresh, unweathered olivine crystals are green or yellow-green in color and look transparent to translucent. Large crystals of green olivine are sometimes used as a gemstone called peridot. However, olivine is a highly reactive mineral and is easily altered. Olivine commonly turns a reddish brown color as a result of partial or complete alteration to iddingsite (a mixture of minerals, including clay minerals and iron oxides).

Olivine also commonly reacts with water to form serpentine minerals, which can also be green in color but which have different textures and lusters than the original olivine. Serpentine minerals often look fibrous and waxy. Sometimes the outside of an olivine crystal will react with water to form serpentine, but a core of fresh olivine will remain. Often, a network of serpentine veins will form in an olivine crystal. The serpentine veins look like a web or mesh netting, and this texture is commonly referred to as "mesh-texture." Olivine can also react with carbon dioxide to form white carbonate minerals, such as magnesite and calcite. Reaction of olivine with carbon dioxide is a natural form of carbon removal from the atmosphere. Scientists are currently studying how this reaction could be increased to reduce carbon dioxide levels in the atmosphere to combat human-caused climate change.

Olivine can be found in rocks such as basalt, peridotite, and serpentinite. However, the olivine in these rocks has often been altered to secondary minerals. The best place to look for fresh green olivine crystals is in young volcanic rocks, such as young basaltic lava flows. In the United States, a good place to look for olivine crystals is in basaltic rocks on the island of Hawaii.

IDENTIFICATION: Unaltered olivine can be recognized by its green or yellow-green color and vitreous luster. In addition, larger crystals of olivine can be recognized by the presence of conchoidal (curved) fracture surfaces. Altered olivine commonly looks reddish brown in color.

FUN FACT: Papakōlea Beach on the island of Hawaii has green sand that is composed of grains of olivine. Green sand beaches are rare, since olivine is an easily weathered mineral and does not survive in most beach environments. However, green sand beaches occur in a few places close to young volcanic rocks that contain olivine crystals.

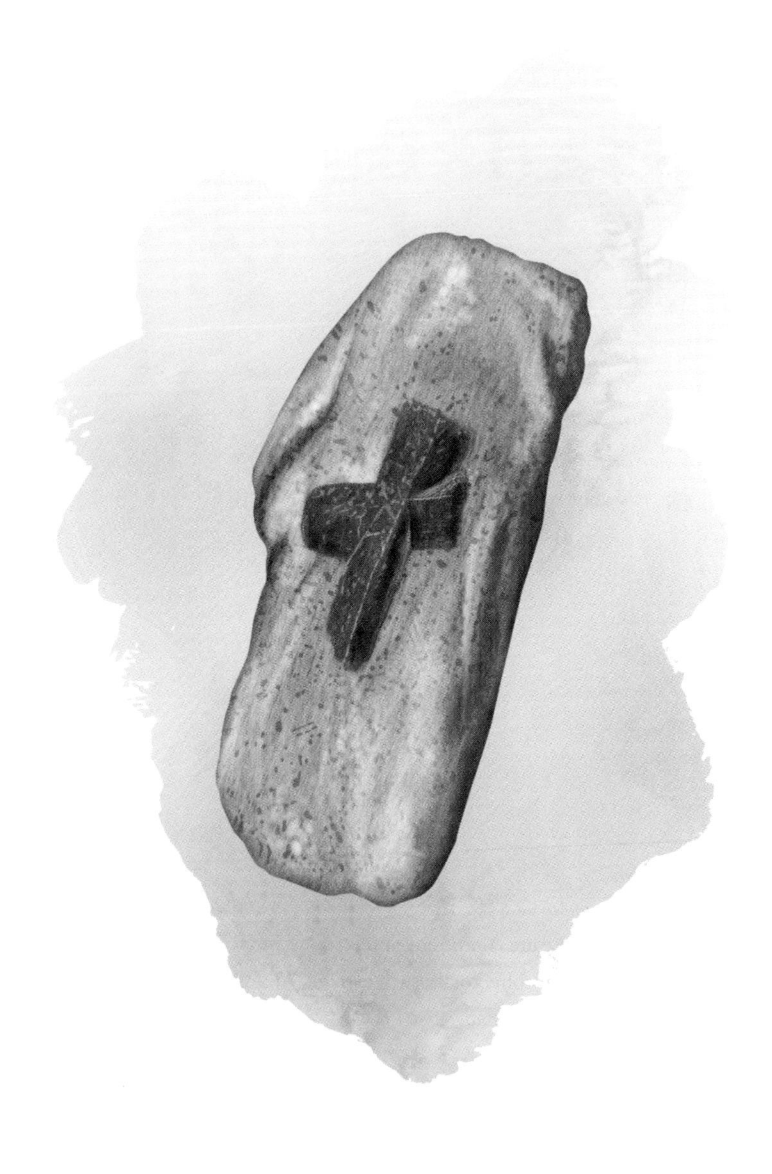

STAUROLITE

FORMULA: $Fe_2Al_9O_6(SiO_4)_4(O,OH)_2$

MINERAL CLASS: Silicate

CRYSTAL SYSTEM: Monoclinic

COLOR: Reddish brown to brownish black

TRANSPARENCY: Translucent to opaque

STREAK: White

LUSTER: Vitreous to resinous

MOHS SCALE HARDNESS: 7.0–7.5

SPECIFIC GRAVITY: 3.7–3.8

DESCRIPTION AND OCCURRENCE: Staurolite is an iron and aluminum silicate mineral that is found in metamorphic rocks. Staurolite is formed under medium- to high-grade pressure and temperature conditions and is found in medium- to high-grade metamorphic rocks, such as schist and gneiss. Staurolite is found in association with other metamorphic minerals, such as muscovite, garnet, and kyanite.

Staurolite is reddish brown or brownish black in color and commonly forms as large, six-sided crystals, which are often twinned. In twinned crystals, two staurolite crystals intersect each other at 60 degree or 90 degree angles, forming a cross shape.

Staurolite can be found in metamorphic rocks around the globe. In the United States, one great place to see staurolite is Fairy Stone State Park in Virginia. The park is named after staurolite, which is colloquially called "fairy stone." Another great place to look for staurolite is Georgia, where staurolite is the official state mineral. Staurolite is particularly common in Fannin County, Georgia, in the Blue Ridge Mountains. Another good place to look for staurolite is New Hampshire, where staurolite can be found in schists located in the western part of the state.

IDENTIFICATION: Staurolite is an easy mineral to identify, especially when twinned. Look for a dark-colored mineral that is shaped like a cross. Even when staurolite is not twinned, it is easy to recognize by its six-sided crystals. Staurolite is usually found within a metamorphic rock, such as schist. However, staurolite crystals can also be found in sediments when they weather out of metamorphic rock.

FUN FACT: The word "staurolite" comes from "stauros," the Greek word for cross.

KYANITE

FORMULA: Al_2SiO_5

MINERAL CLASS: Silicate

CRYSTAL SYSTEM: Triclinic

COLOR: Usually blue

TRANSPARENCY: Transparent to translucent

STREAK: White

LUSTER: Usually vitreous, sometimes pearly

MOHS SCALE HARDNESS: Variable—5 along length of crystals, 7 across crystals

SPECIFIC GRAVITY: 3.5–3.7

DESCRIPTION AND OCCURRENCE: Kyanite is an aluminum silicate mineral that is usually found in metamorphic rocks, most commonly in schists and gneisses. Kyanite is also sometimes found in eclogites and pegmatites.

Kyanite generally forms distinctive long, bladed crystals and is most commonly a light blue color, although it sometimes appears more blue-gray than blue. Less commonly, kyanite can be other colors, such as white, green, orange, and black. A distinctive property of kyanite is that it has different hardness in different directions. Along the length of crystals, kyanite has a hardness of about 5. However, across crystals (at 90 degree angles to the length), kyanite has a hardness of about 7.

Kyanite is a well-known example of a mineral that has polymorphs, which are minerals that have identical chemical compositions but different crystal structures. Polymorphs generally occur as a result of different pressure and temperature conditions. The polymorphs of kyanite are andalusite and sillimanite. The kyanite polymorph forms when pressure is high and temperature is relatively low. Under lower pressure conditions, the andalusite polymorph will form. Under higher temperature conditions, the sillimanite polymorph will form.

Kyanite is mined for use in various industries, for example in the production of porcelain and spark plugs for vehicles. The world's largest and oldest kyanite mine is located in Virginia, United States, where kyanite has been mined since the 1940s.

IDENTIFICATION: Kyanite can be identified by its blue color (although bear in mind it is not always blue) and long, blade-shaped crystals. A good way to identify kyanite is to test its hardness, since it is the only common mineral with a detectable difference in hardness in different directions. A knife blade or steel needle should scratch kyanite along the length of the crystals but should not scratch kyanite across the crystals.

FUN FACT: An old name for kyanite is "disthene", which means "two strengths" in ancient Greek. The modern word "kyanite" comes from the ancient Greek word "kyanos", which means blue.

ZIRCON

FORMULA: $ZrSiO_4$

MINERAL CLASS: Silicate

CRYSTAL SYSTEM: Tetragonal

COLOR: Variable—often brown or reddish brown

TRANSPARENCY: Variable

STREAK: White

LUSTER: Vitreous to adamantine

MOHS SCALE HARDNESS: 7.5

SPECIFIC GRAVITY: 4.6–4.7

DESCRIPTION AND OCCURRENCE: Zircon is a zirconium silicate mineral that is commonly found as a minor component of igneous rocks, such as granites and pegmatites. Zircon is more common in felsic (silica-rich) igneous rocks and is less abundant or absent in mafic (silica-poor) igneous rocks. Zircon can also be found in many metamorphic rocks, such as marbles, schists, and gneisses. Zircon is an extremely durable mineral, so it is commonly found in sediments and sedimentary rocks. Zircon that is found in sediments and sedimentary rocks is known as "detrital zircon."

Zircon mineral grains are usually small, generally less than 1 millimeter in size. Thus, zircon is usually observed using a microscope. In some pegmatites and in some metamorphic rocks, zircon crystals can be much larger, up to several centimeters in size. Zircon is most commonly brown or reddish brown in color. However, it can also be colorless and other colors, such as yellow, green, gray, and red. When zircon forms crystals, the crystals are four-sided prisms that have a pyramid on each end.

Zircon is a hard mineral and is mined for use as an abrasive. In addition, zircon is mined to produce the metal zirconium, which is used in ceramics and also to create corrosion-resistant alloys that are used in special applications, for example in pipes for chemicals and in cladding for nuclear reactors. Most zircon mining takes place in Australia and South Africa. In the United States, zircon is mined from heavy mineral sands in Florida and Virginia. Zircon is also used as a gemstone. Colorless zircon looks similar to diamond, although it is less resistant to scratching. Colored zircons, such as red, pink, green, and yellow zircons, are also used as gemstones.

Zircon is an important mineral for geologists because it can be used to determine the ages of rocks. Zircon crystals contain a small amount of uranium, which is radioactive and transforms to lead over time. Scientists can measure the proportions of uranium and lead in zircons to determine the ages of the mineral grains.

IDENTIFICATION: Because zircon grains are typically small, they are difficult to see in most rocks. Geologists often look for zircon by preparing a thin slice of rock and examining it under a microscope.

When geologists want to extract zircons for dating a rock, they generally crush the rock and use special magnetic and density equipment to separate zircons from the other minerals. When zircon can be observed, it can be recognized by its crystal shape (four-sided prisms with pyramids on both ends) and its high specific gravity. In addition, zircon usually has a brown or reddish brown color.

FUN FACT: Zircons are the oldest minerals on Earth. In the Jack Hills region of Western Australia, zircons more than 4 billion years old have been found. These zircons are detrital zircons, which means that they weathered out of older rocks (now completely gone) and were later incorporated into younger rocks. The oldest Jack Hills zircon analyzed to date has an age of 4.4 billion years!

BERYL

FORMULA: $Be_3Al_2Si_6O_{18}$

MINERAL CLASS: Silicate

CRYSTAL SYSTEM: Hexagonal

COLOR: Variable

TRANSPARENCY: Transparent to translucent

STREAK: White

LUSTER: Vitreous

MOHS SCALE HARDNESS: 7.5–8.0

SPECIFIC GRAVITY: 2.6–2.8

DESCRIPTION AND OCCURRENCE: Beryl is a beryllium and aluminum silicate mineral that is primarily found in igneous and metamorphic rocks. Beryl is most commonly found in granitic rocks, particularly in granitic pegmatites. In some pegmatites, very large beryl crystals several meters in length have been found. Beryl is also found in mica-rich schists and gneisses. Beryl can also form through hydrothermal alteration of sedimentary rocks, most notably in limestone and shale rocks in Colombia.

When beryl crystals form, for example in a pegmatite, they are usually hexagonal in shape. Beryl comes in a wide variety of colors, including green, blue, pink, red, yellow, and colorless. Beryl is a well-known gemstone. Different colors of beryl gemstone are known by different names, including emerald (green), aquamarine (blue), morganite (pink), bixbite (red), heliodor (yellow), and goshenite (colorless). The different colors of beryl are caused by the presence of different trace elements, for example chromium and vanadium in emerald and iron in aquamarine.

Beryl can be found in igneous and metamorphic rocks all over the globe. In the United States, beryl can be found in a few regions, including New England as well as the states of North Carolina, Colorado, Utah, Idaho, and California.

Gemstone varieties of beryl are found all over the world, although a few countries dominate gemstone production. Most emerald production comes from deposits in Colombia. In addition, Brazil, Zambia, Zimbabwe, and Russia are major producers of emeralds. The most important producer of other beryl gemstones, such as aquamarine and morganite, is Brazil. In the United States, gem beryl deposits can be found in places such as North Carolina (emerald), Colorado (aquamarine), New Hampshire (goshenite), and Utah (bixbite).

In addition to being mined as a gemstone, beryl is mined for the element beryllium. Historically, beryl has been mined for beryllium in countries such as Brazil, Madagascar, Mozambique, Russia, China, and the United States. Today, the United States dominates beryllium production, but this beryllium is mostly produced from bertrandite (another beryllium-rich mineral) rather than beryl.

In the past, beryl was mined for beryllium in the states of New Hampshire, Massachusetts, Connecticut, and Maine.

IDENTIFICATION: When crystals form, beryl can be easily recognized by its hexagonal-shaped crystals. Apatite also forms hexagonal crystals and can look similar to beryl, but it is much softer (hardness of 5). Massive beryl could be mistaken for quartz but is slightly harder and cannot be scratched by quartz.

FUN FACT: Beryl was a popular girl's name in the late 1800s and first few decades of the 1900s, particularly in England. Famous women named Beryl include Beryl Markham, a pioneering aviator; Beryl Bainbridge, an award-winning author; and Beryl Cook, an artist.

CALCITE

FORMULA: $CaCO_3$

MINERAL CLASS: Carbonate

CRYSTAL SYSTEM: Trigonal

COLOR: Usually colorless or white

TRANSPARENCY: Variable

STREAK: White

LUSTER: Usually vitreous

MOHS SCALE HARDNESS: 3

SPECIFIC GRAVITY: 2.6–2.8

DESCRIPTION AND OCCURRENCE: Calcite is a very common carbonate mineral and the most stable of the three polymorphs (minerals with the same chemical formula but different crystal structures) of $CaCO_3$. The other two polymorphs are aragonite, which is often formed through biological processes, and vaterite, which can be found near mineral springs. Aragonite and vaterite are metastable under most conditions and will often convert to calcite over time.

Calcite is very similar to a magnesium-bearing mineral called dolomite ($CaMg(CO_3)_2$). Like aragonite, calcite can form through biological processes. Calcite can be found in the shells of marine organisms such as plankton, algae, and some bivalves. Calcite also commonly forms through inorganic precipitation from water, such as seawater or hot spring fluids.

Calcite is most commonly colorless or white in color. However, it can also be other colors, including gray, yellow, green, blue, red, and brown. Calcite has three cleavage planes (surfaces along which a mineral tends to break) that create a rhombohedral shape when calcite breaks along these planes. Thus, crystals of calcite will often have a rhombohedral shape. Clear calcite displays an unusual mineral property: birefringence, or the double refraction of light. When clear calcite crystals are placed on top of an image, such as lines of text, the image will be doubled. Clear calcite that is birefringent is often called "Iceland spar," because this is where this type of calcite was originally found. Calcite will vigorously react with acid, such as hydrochloric acid or vinegar, releasing carbon dioxide gas bubbles.

Calcite is found in many sedimentary rocks, such as limestone. Calcite can be found in stalactites and stalagmites that form in limestone caves. Calcite is also found in carbonate-rich metamorphic rocks, such as marble. Calcite is a common secondary (replacement) mineral in igneous rocks. For example, calcite is often found as a secondary mineral filling vesicles (holes) in basalt. Rarely, calcite can form as a primary igneous mineral, for example in carbonatites. Calcite can be found near hot springs, where it precipitates from heated waters to form travertine. Less commonly, calcite precipitates from cooler waters to form tufa.

In the United States, calcite can be found just about anywhere in many different rock types. If you want to add a sample of Iceland spar to your collection, one place where you can find it is the Harding Pegmatite Mine in New Mexico.

IDENTIFICATION: A good way to identify calcite is its tendency to break into blocks that have a rhombohedral shape. Clear crystals of calcite can display birefringence, which will cause an image to look doubled when viewed through the calcite crystal. Calcite can look similar to quartz, dolomite, gypsum, and halite. Calcite is softer than quartz (hardness of 7) and harder than gypsum (hardness of 2). You will be able to scratch gypsum with your fingernail (hardness of 2.5) but not calcite. In addition, calcite with fizz when exposed to dilute acid, such as vinegar, whereas quartz and gypsum will not fizz. Dolomite will fizz when exposed to acid, but it fizzes less vigorously than calcite. The difference will be more obvious with hydrochloric acid compared to vinegar, but you must take appropriate safety precautions when using hydrochloric acid. Halite can be easily identified by its salty taste.

FUN FACT: Iceland spar, a type of clear calcite, was used in the 1800s and first half of the 1900s to create polarizing prisms that were used in microscopes and other scientific instruments. Today, most scientific instruments use other types of polarizers, such as plastic polaroid sheets.

MALACHITE

FORMULA: $Cu_2CO_3(OH)_2$

MINERAL CLASS: Carbonate

CRYSTAL SYSTEM: Monoclinic

COLOR: Bright green

TRANSPARENCY: Translucent to opaque

STREAK: Pale green

LUSTER: Variable

MOHS SCALE HARDNESS: 3.5–4.0

SPECIFIC GRAVITY: 3.5–4.0

DESCRIPTION AND OCCURRENCE: Malachite is a bright green, copper-rich carbonate mineral that commonly forms through the weathering of copper ore bodies. Typically, malachite is formed when copper ore bodies are located in the proximity of limestone rock, which provides the carbonate that goes into malachite's mineral structure. Malachite is often found in association with blue azurite, another copper-rich carbonate mineral. Malachite is also commonly found with calcite, another carbonate mineral, and with other copper-bearing minerals, such as bornite, chalcopyrite, cuprite, and native copper.

Malachite is usually bright green in color, commonly containing bands of alternating lighter and darker green colors. Malachite is most commonly found in massive form with a botryoidal (grape-like) morphology. Sometimes, malachite forms inside limestone caves as stalactites and stalagmites. When malachite crystals form, they are acicular (needle-like) or fibrous in shape. Massive malachite generally has a dull luster. Acicular crystals have a vitreous to adamantine luster while fibrous crystals have a silky luster.

In the United States, malachite is is found in copper-rich carbonate rocks in the Southwest, most notably in Arizona. Sometimes these rocks are mined for copper—for example, at Morenci Mine in Arizona.

IDENTIFICATION: Malachite is easily recognized by its bright green color. Malachite is a soft mineral and can be easily scratched with a knife. Malachite is often found in association with blue azurite. Because it is a carbonate mineral, malachite will fizz when exposed to dilute acid, such as vinegar.

FUN FACT: Malachite has been used as a pigment for green paint since ancient times. For example, the ancient Egyptians used malachite paint to decorate their tombs and also used crushed up malachite as green eye make-up. Today, green paint is generally made using artificial pigments.

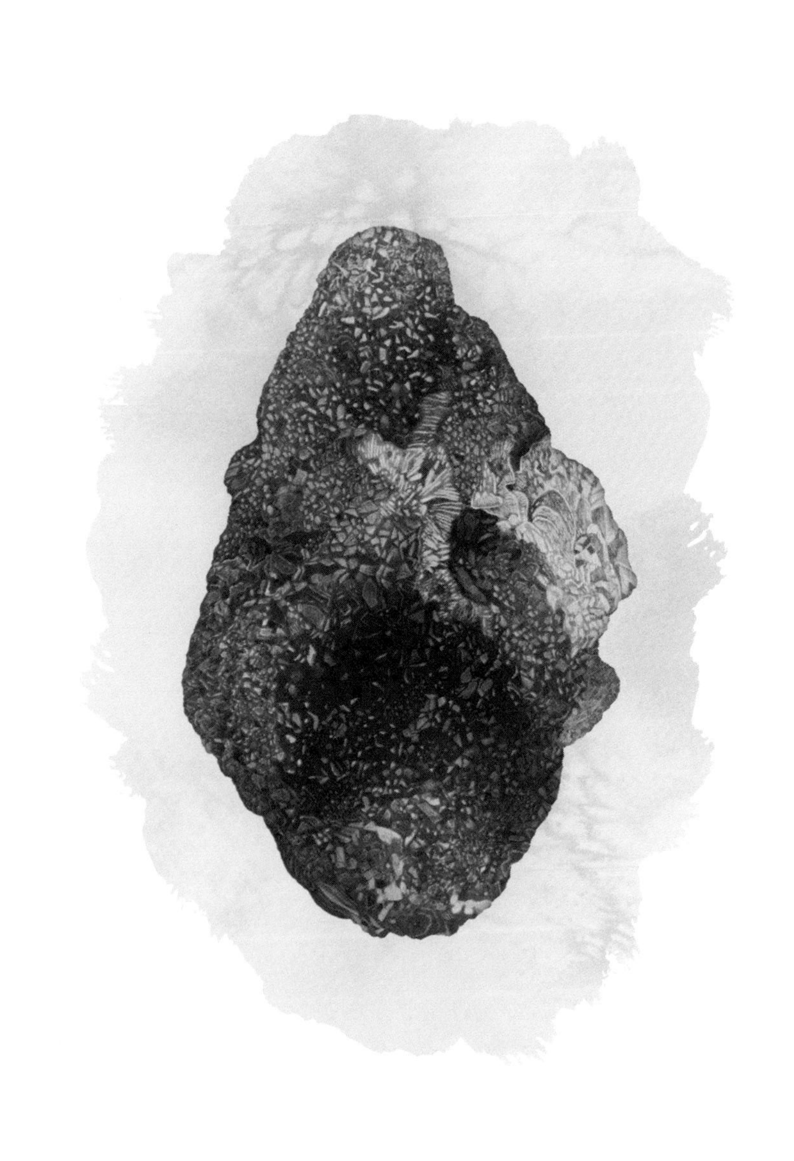

AZURITE

FORMULA: $Cu_3(CO_3)_2(OH)_2$

MINERAL CLASS: Carbonate

CRYSTAL SYSTEM: Monoclinic

COLOR: Deep blue

TRANSPARENCY: Transparent to opaque

STREAK: Light blue

LUSTER: Vitreous

MOHS SCALE HARDNESS: 3.5–4.0

SPECIFIC GRAVITY: 3.7–3.9

DESCRIPTION AND OCCURRENCE: Similar to malachite, azurite is a copper-rich carbonate mineral that commonly forms through the weathering of copper ore bodies located in the vicinity of limestone rocks. However, azurite is a deep blue color while malachite is a bright green color. Azurite is commonly found in association with malachite. Azurite is less stable than malachite and will actually transform into malachite over time, through a chemical weathering reaction that adds water to the mineral structure and releases carbon dioxide. Like malachite, azurite is found in association with calcite and other copper-bearing minerals.

Azurite is most commonly found in massive form, although crystals can also form in limestone cavities. When crystals form, they are usually prismatic (shaped like a prism) or tabular in shape. Like malachite, azurite can have a botryoidal (grape-like) morphology and can also form as stalactites and stalagmites in caves.

In the United States, azurite is found in the same copper-rich rocks that contain malachite. These rocks are mostly located in the Southwest. A famous mineral collecting location for azurite is The Lavender Pit, which is located inside an old copper mine in Bisbee, Arizona. Many spectacular azurite specimens have been collected from this pit, including some that have a bright, electric blue color.

IDENTIFICATION: Azurite is easily recognized by its striking deep blue color and common association with green malachite. Azurite is a soft mineral and can be easily scratched with a knife. Because it is a carbonate mineral, azurite will fizz when exposed to dilute acid, such as vinegar.

FUN FACT: Azurite is sometimes called "Chessylite," since the mineral was first described from an occurrence near Chessy-les-Mines, near Lyon, France.

CORUNDUM

FORMULA: Al_2O_3

MINERAL CLASS: Oxide

CRYSTAL SYSTEM: Trigonal

COLOR: Variable

TRANSPARENCY: Transparent to opaque

STREAK: White

LUSTER: Adamantine to vitreous

MOHS SCALE HARDNESS: 9

SPECIFIC GRAVITY: 3.9–4.1

DESCRIPTION AND OCCURRENCE: Corundum is an aluminum oxide mineral that can be found in both igneous and metamorphic rocks. In addition, because corundum is a very hard and durable mineral, it can be found in sediments. In most rocks, corundum is a minor mineral component of the rock. However, a metamorphic rock called emery consists mostly of corundum, along with other iron-rich minerals, such as magnetite and hematite. In the past, emery was commonly used to manufacture abrasive material, such as "emery paper."

A distinguishing feature of corundum is that it is one of the hardest minerals known. It is harder than all other common minerals, although diamond (a rare mineral) is harder than corundum. Corundum often forms as barrel-shaped or spindle-shaped crystals.

Corundum is found in many different colors, including brown, red, blue, pink, purple, orange, green, and yellow. Corundum is famous for its use as a gemstone, and different color corundum gemstones are known by different names. Gemstone quality corundum that is red in color is called ruby. Gemstone quality corundum that is another color is called sapphire. Sapphires are most commonly blue, but they can also be other colors, such as pink, orange, yellow, green, and purple. The different colors of corundum are caused by different trace elements. For example, chromium causes the red color in rubies and iron and titanium cause the blue color in blue sapphires.

Corundum can be found in many rocks around the world. Gemstone quality rubies and sapphires are usually found in alluvial (stream) sediments, after they have weathered out of metamorphic rocks. Historically, most rubies and sapphires have been mined in Australia and in Asia, for example in Myanmar, Sri Lanka, Thailand, and Cambodia. In the United States, blue sapphires have sporadically been mined at Yogo Gulch, Montana, since the late 1800s. Sapphires also occur in stream and river sediments in western Montana, including in the Missouri River. Rubies and sapphires can also be found in the mountains of western North Carolina, particularly in Macon County, where the general public can visit local mines and search for gemstones for a fee.

IDENTIFICATION: Corundum can be identified by its hardness and relatively high specific gravity. The presence of barrel-shaped or spindle-shaped crystals is another good way to identify corundum. Identification of corundum in gemstones, especially those set in jewelry, is best done at a reputable jewelry shop.

FUN FACT: The Liberty Bell Ruby is a statue that was carved into one of the world's largest rubies. The ruby was mined in East Africa in the 1950s and weighs more than four pounds! The ruby was carved to look like the famous Liberty Bell in Philadelphia and decorated with 50 diamonds, which represent the U.S. states. The Liberty Bell Ruby was stolen from a jewelry store in Delaware in 2011 and has never been recovered by police.

MAGNETITE

FORMULA: Fe_3O_4

MINERAL CLASS: Oxide

CRYSTAL SYSTEM: Cubic

COLOR: Black

TRANSPARENCY: Opaque

STREAK: Black or dark silvery gray

LUSTER: Metallic

MOHS SCALE HARDNESS: 5.5–6.5

SPECIFIC GRAVITY: 5.2

DESCRIPTION AND OCCURRENCE: Magnetite is a common oxide mineral that is very rich in iron and has a closely packed mineral structure, which makes it a heavy mineral. Magnetite is an important ore mineral for iron. Magnetite is a magnetic mineral and is strongly attracted to a magnet. Rarely, magnetite can become naturally magnetized, which means that it will attract pieces of iron metal. Geologists are not completely sure how this natural magnetization occurs, but it is potentially caused by the natural magnetic fields that surround lightning bolts. Naturally magnetized magnetite is called "lodestone" and was used in early magnetic compasses.

Magnetite is either dark black or dark silvery gray in color and has a metallic luster. Magnetite is often massive. When crystals form, they are usually octahedral in shape. Less commonly, the crystals are dodecahedral in shape.

Magnetite occurs in small quantities in igneous rocks. In some igneous rocks, magnetite can become concentrated through magmatic processes. Magnetite is also found in both metamorphic and sedimentary rocks. Magnetite is a durable mineral and can survive weathering processes that break down other minerals. For this reason, magnetite can be found in sediments and sedimentary rocks. Magnetite is sometimes mined from beach sand.

Magnetite is found in an interesting sedimentary rock known as banded iron formation, which consists of alternating layers of iron-rich magnetite or hematite (another iron oxide mineral) and iron-poor chert (a type of quartz). Banded iron formation rocks are mostly very old. Most were formed in Archean times more than 2.5 billion years ago! The formation of the oxide minerals found in banded iron deposits is related to the increased presence of oxygen on Earth due to the evolution of life. Thus, geologists are very interested in studying these rocks. In addition, banded iron formation rocks are mined as an important source of iron.

Magnetite can be found in many rocks and sediments throughout the United States. One good place to look for magnetite is on a beach, where heavy minerals (including magnetite) can be concentrated through wave action. Often, the dark concentrations of minerals that you see on an otherwise white beach contain a significant amount of magnetite. Magnetite can also be found in banded iron formation rocks located in Minnesota, Wisconsin, and Michigan near Lake Superior. The lodestone variety of magnetite can be found near the aptly named town of Magnet Cove, Arkansas.

IDENTIFICATION: Magnetite is magnetic, so that is an easy way to identify it! A large crystal of magnetite will strongly attract a magnet. When small crystals of magnetite are present in a rock, the rock may also be magnetic, although the effect may be less strong. For example, a magnet might not stick to a rock with small magnetite crystals, but a magnet hanging from a string may be pulled towards the rock. Magnetite can also be identified by its dark color and high specific gravity.

FUN FACT: Magnetite played a key role in the discovery of plate tectonics. The Earth's magnetic field has switched polarity (the north pole became the south pole, and vice versa) many times in the past. When volcanic rocks that contain magnetite cool and crystallize, they become magnetized in the same direction as the magnetic field at the time that they cool. In the 1960s, scientists studying volcanic rocks on the seafloor realized that the rocks had alternating directions of magnetism, forming striped patterns on either side of a volcanic ridge. Scientists realized that the rocks must have formed at the ridge and then moved away from it over time, which provided concrete evidence that Earth's crust moves.

PYRITE

FORMULA: FeS_2

MINERAL CLASS: Sulfide

CRYSTAL SYSTEM: Cubic

COLOR: Usually pale brass yellow

TRANSPARENCY: Opaque

STREAK: Greenish black or brownish black

LUSTER: Metallic

MOHS SCALE HARDNESS: 6.0–6.5

SPECIFIC GRAVITY: 4.9–5.2

DESCRIPTION AND OCCURRENCE: Pyrite is the most common sulfide mineral and is found worldwide in many different kinds of igneous, sedimentary, and metamorphic rocks. Pyrite has a metallic luster and is usually a pale brass yellow color. However, when pyrite tarnishes it can look darker. When crystals form, they are usually cubic or octahedral in shape. Pyrite also commonly forms as rounded balls known as "nodules." In some sedimentary rocks, pyrite forms as flat discs, which are called "pyrite suns" or "sun dollars."

Pyrite is found in many different types of igneous rocks, including granite and basalt. Pyrite is more common in igneous rocks that are rich in sulfur. Pyrite is also commonly found in sedimentary rocks, particularly in mudstone, coal, and limestone. Pyrite is also found in metamorphic rocks, including slate, schist, and gneiss. Pyrite is a common mineral in hydrothermal veins, which form as a result of the circulation of hot fluids in rocks. In hydrothermal veins pyrite is often found in association with ore deposits, for example deposits of gold, copper, or silver.

In the United States, pyrite can be found in many different rocks and in many mountain streams across the country. Pyrite crystals are easy to purchase in rock shops or museum gift shops, although the specimens are likely sourced from international locations such as Peru, Spain, or Mexico. The pyrite sun specimens available for sale in rock shops generally originate from Sparta, Illinois, where they are found in underground coal mines. Pyritized fossils, which are golden-colored fossils that contain pyrite, are particularly beautiful specimens to collect. Some great examples of pyritized fossils are found in the Silica Shale in Ohio.

IDENTIFICATION: Pyrite is easily identified by its pale brass yellow color, metallic luster, dark streak, and high specific gravity. Pyrite crystals can be identified by their cubic or octahedral shapes. Pyrite looks similar to chalcopyrite, but chalcopyrite is softer (hardness of 3.5 to 4.0). In addition, when chalcopyrite weathers it develops a characteristic iridescence that makes it look different from pyrite. Pyrite can look superficially similar to gold, but gold is much softer (hardness of 2.5 to 3.0) and heavier (specific gravity of about 19). In addition, gold usually looks brighter than pyrite.

FUN FACT: Pyrite is also known as "Fool's Gold" due to its superficial resemblance to gold. It fooled many miners during the California Gold Rush. Don't let it fool you!

GALENA

FORMULA: PbS

MINERAL CLASS: Sulfide

CRYSTAL SYSTEM: Cubic

COLOR: Silver or dark gray

TRANSPARENCY: Opaque

STREAK: Dark gray to black

LUSTER: Metallic

MOHS SCALE HARDNESS: 2.5

SPECIFIC GRAVITY: 7.4–7.6

DESCRIPTION AND OCCURRENCE: Galena is one of the most common sulfide minerals and is the primary ore for lead. Galena is often found in lead-zinc deposits that can also contain silver, copper, antimony, and bismuth. Galena can be found in hydrothermal veins (formed when hot fluids circulate through rocks) that form in a number of different rock types and in pegmatites (a type of igneous rock with very large crystals). Large deposits of galena can be found in sedimentary rocks (especially limestone) that have experienced hydrothermal alteration.

Galena has a characteristic metallic luster and is silver or dark gray in color. When galena crystals form, they are usually cubic or octahedral in shape. Galena can also be found in massive form, with no visible crystals. Galena often has a cubic shape, either from initial crystal growth or because it has been broken into cubes. Galena has three planes of cleavage (surfaces along which a mineral tends to break) that intersect at 90 degree angles and thus it tends to break into cubes.

In the United States, large amounts of galena are located in the "lead belt" ore deposits in southeastern Missouri, where lead has been mined from the 1700s to the present day. The prevalence of these deposits throughout the Mississippi River valley has been used to describe similar deposits worldwide as "Mississippi Valley Type Deposits." In addition to the southeastern Missouri lead district, the tri-state district of southwestern Missouri, southeastern Kansas, and northwestern Oklahoma was also mined for galena for over a hundred years. At its peak, this region was one of the world's largest lead producing regions and was a strategic source of lead for the United States and Allied forces during World War I and World War II. Galena also occurs in many other mining districts in the United States, including in Alaska, Nevada, Illinois, Iowa, Wisconsin, and Tennessee.

IDENTIFICATION: Galena is easily recognized by its silver or dark gray color, metallic luster, cubic shape, softness, and high specific gravity. Note that galena generally has a bright silver color on fresh surfaces but can tarnish to a dull gray color.

FUN FACT: The states of Kansas, Missouri, Indiana, and Illinois all have towns named Galena. In the past, large amounts of galena ore were mined in, around, or beneath these towns.

GYPSUM

FORMULA: $CaSO_4 \cdot 2H_2O$

MINERAL CLASS: Sulfate

CRYSTAL SYSTEM: Monoclinic

COLOR: Commonly colorless to white

TRANSPARENCY: Transparent to opaque

STREAK: White

LUSTER: Vitreous, silky, or pearly

MOHS SCALE HARDNESS: 2

SPECIFIC GRAVITY: 2.3

DESCRIPTION AND OCCURRENCE: Gypsum is a common sulfate mineral that commonly crystallizes by evaporation of sulfate-rich waters, such as seawater, ephemeral lakes, and groundwater in contact with coals or pyrite-bearing mudstones. Thick beds of gypsum can be found in sedimentary rock units in association with other evaporite minerals, such as halite and anhydrite. In addition to forming directly through evaporation, gypsum also commonly forms through the hydration of anhydrite, which has the chemical formula $CaSO_4$ (the same formula as gypsum, but with no water). In addition, gypsum can form from sulfur-bearing hydrothermal fluids (hot fluids) associated with volcanic activity.

The appearance of gypsum is variable. Gypsum is most commonly colorless to white in color, but it can also be many other colors, including yellow, pink, red, gray, brown, and tan. When gypsum is colorless and transparent, it is called selenite. This type of gypsum has a vitreous or pearly luster. Another type of gypsum called satin spar has a fibrous texture and silky luster. Alabaster is a fine-grained, massive variety of gypsum. In arid desert environments, something called a "gypsum rose" or "desert rose" can form. These roses have mineral crystals that look like the petals of a flower and are generally composed of gypsum and sand grains. Baryte, another sulfate mineral, can also form similar rose-like shapes, but they are less delicate in form. A gypsum rose makes a beautiful addition to a rock collection and can be purchased in most rock shops as well as in many museum gift shops.

Gypsum has many uses and is commonly mined. Gypsum is used in the construction industry. For example, it is the major component in drywall and is used as a raw material in the production of some cements. Gypsum is also used in the production of plaster and is commonly used as a fertilizer. The alabaster variety of gypsum is commonly used as an ornamental carving stone.

In the United States, large deposits of gypsum are found in several states, including California, Colorado, Texas, Oklahoma, Iowa, and New York. Gypsum sand is rare because gypsum will dissolve in water over time. However, extraordinary large dunes of gypsum sand are located in White Sands National Park in New Mexico, which protects this unusual sand formation from commercial exploitation. Another great place to see gypsum is Carlsbad Caverns in New Mexico, where several cave chambers contain large blocks of gypsum.

IDENTIFICATION: When identifying gypsum, bear in mind that it varies widely in color, transparency, and morphology. A good way to identify gypsum is through its softness. Gypsum has a hardness of 2 and can be scratched by a fingernail, which has a hardness of 2.5.

FUN FACT: The term "selenite," which is used to refer to colorless, transparent crystals of gypsum, means "moonstone." The word originates from "selene," the ancient Greek word for moon. However, be careful not to confuse selenite with the gemstone moonstone, which is a type of feldspar.

HALITE

FORMULA: NaCl

MINERAL CLASS: Halide

CRYSTAL SYSTEM: Cubic

COLOR: Variable—commonly colorless or white

TRANSPARENCY: Transparent to translucent

STREAK: White

LUSTER: Vitreous

MOHS SCALE HARDNESS: 2.5

SPECIFIC GRAVITY: 2.1–2.2

DESCRIPTION AND OCCURRENCE: Halite is also known as rock salt. Halite is very soft, has a vitreous luster, and is transparent to translucent. When crystals form, they are usually cubic in shape. However, halite can also be found in massive form, with no visible crystals. Halite is most commonly colorless or white, but it can also be pink, red, yellow, blue, or gray in color.

Halite is a common mineral and can be found anywhere that salty water has evaporated. For example, you can sometimes find halite crystals in the desert where lakes have evaporated. Large deposits of halite are generally found where seawater or inland salty lakes evaporated in the geologic past. These deposits can sometimes be hundreds of meters thick. Halite most often occurs in association with other evaporate minerals, such as gypsum, anhydrite, and sylvite. Sometimes, deformation of halite deposits leads to the formation of large salt domes. Salt domes are important in the fossil fuel industry because they can act as a trap for oil and gas. In the United States, large deposits of halite are found in many states, such as New York, Michigan, Ohio, Kansas, Oklahoma, Texas, and Louisiana.

IDENTIFICATION: Taste is not a common method for identifying minerals. However, it is a perfect way to identify halite, which tastes salty. Halite can also be identified by its cubic shape and the fact that it easily dissolves in water.

FUN FACT: Most table salt is produced from halite deposits. Natural halite deposits are mined by dissolving the halite deep beneath the surface, forming a brine that is then brought to the surface, purified, and evaporated in shallow pools. This purified halite is the white salt commonly found in everyone's kitchen.

RARE MINERALS

DIAMOND

FORMULA: C

MINERAL CLASS: Native element

CRYSTAL SYSTEM: Cubic

COLOR: Usually colorless or white

TRANSPARENCY: Usually transparent

STREAK: Colorless

LUSTER: Adamantine

MOHS SCALE HARDNESS: 10

SPECIFIC GRAVITY: 3.5

DESCRIPTION AND OCCURRENCE: Diamond is a rare mineral that is made up of the element carbon. The carbon atoms are closely packed together and are strongly bonded to each other, which makes diamond one of the hardest naturally occurring minerals. Natural diamond only forms under high pressure and high temperature conditions located deep within Earth's mantle. Closer to Earth's surface, carbon crystallizes as the mineral graphite, which has a different crystal structure and is very soft.

Diamonds are most commonly colorless or white. However, they can also be yellow, blue, green, orange, pink, purple, red, gray, brown, and black in color. The different colors are generally the result of trace elements or deformation of the diamond structure. For example, blue diamonds result from trace amounts of boron while yellow diamonds usually result from trace amounts of nitrogen. Pink and brown diamonds are believed to result from deformation of the diamond structure. Some green diamonds obtained their color due to irradiation, most likely from adjacent radioactive minerals.

Most diamonds are transparent to translucent, although some are opaque. Diamond crystals commonly have an octahedral shape (looks like two pyramids joined at their bases). More rarely, diamond crystals can form in another shape, such as a cube or dodecahedron. Diamonds have a very reflective adamantine luster. Diamond gemstones are generally measured in carats, with one carat being equivalent to 0.2 grams. Only some diamonds are gem quality and suitable for use in jewelry. Lower quality diamonds, which are sometimes described using the term "bort," are generally used in abrasives. Synthetic (lab-grown) diamonds are also manufactured for use in abrasives. Increasingly, synthetic diamonds are also being manufactured for use in jewelry.

Natural diamonds are formed in Earth's mantle, usually at depths of about 150 to 200 kilometers. Most diamonds are very old and were formed billions of years ago. At Earth's surface, diamond can be found in kimberlites and lamproites, which are rare ultramafic (very silica-poor) volcanic rocks that originate from melting deep within the Earth. When kimberlite and lamproite magmas pass through a part of Earth's mantle that contains diamonds, they can bring the diamonds up to Earth's surface as xenocrysts, or "foreign crystals."

However, diamonds are not found in all kimberlite and lamproite rocks, and economic deposits of diamonds are very rare. In addition, diamonds can be found in placer deposits, which are located in sediments in rivers, beaches, and the ocean. The diamonds found in placer deposits weathered out of their original volcanic rocks. Placer deposits generally form when diamonds travel down rivers and then concentrate in gravels in a high-energy environment, for example on a beach.

Most diamonds come from Africa, where they are mined in several countries, including South Africa, Namibia, Botswana, Angola, Zimbabwe, Lesotho, Sierra Leone, and the Democratic Republic of the Congo. Other countries known for diamond production are Russia, India, Brazil, and Canada. For more than thirty years, diamonds were also produced from a mine called Argyle in Western Australia, which closed in 2020. Some diamond mines are known to produce colored stones. For example, the Argyle Mine produced a large number of brown and pink diamonds, and many beautiful blue diamonds have come from Cullinan Mine in South Africa.

In the United States, diamonds can be found in kimberlite and lamproite pipes in a few states, such as Arkansas, Colorado, and Wyoming. However, diamond deposits in the United States are generally not economic to mine. From 1996 to 2001, diamonds were mined in Colorado at the Kelsey Lake Diamond Mine. In the early 1900s, diamonds were mined near Murfreesboro, Arkansas. The Arkansas mine has now become Crater of Diamonds State Park. Today, there is no commercial diamond mining at this location. However, the general public can visit the park and pay a small fee to prospect for diamonds. Every year, visitors find hundreds of diamonds at the site, although most are small and not very valuable.

IDENTIFICATION: Diamond can be identified by its hardness. Uncut diamond crystals are often octahedral in shape. Diamond is most commonly colorless or white but can also be many other colors, including yellow, blue, and pink. Synthetic (lab-grown) diamonds look virtually identical to natural diamonds. In order to tell if a diamond is synthetic or not, you will need to visit a gem expert who has specialist equipment. When purchasing a diamond gemstone, it is a good idea to ask if the stone has certification paperwork—for example, documentation issued by the Gemological Institute of America.

FUN FACT: The largest diamond ever found in the United States is the Uncle Sam Diamond, which was found in Arkansas in 1924. Originally just over 40 carats in size, the diamond was cut into a 12.42 carat emerald-shaped gemstone, which is on display at the Smithsonian Institution in Washington, D.C.

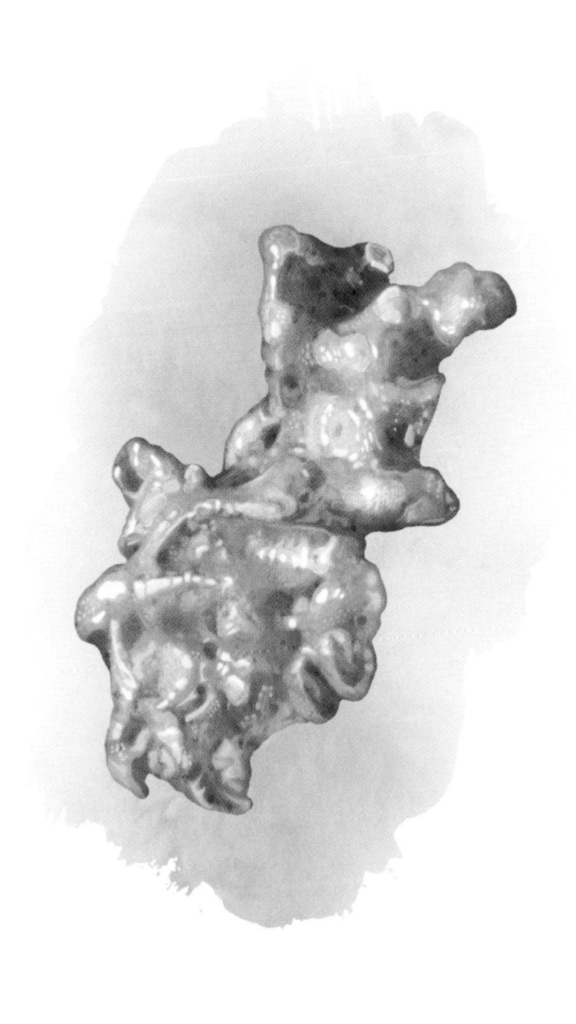

GOLD

FORMULA: Au

MINERAL CLASS: Native element

CRYSTAL SYSTEM: Cubic

COLOR: Golden yellow

TRANSPARENCY: Opaque

STREAK: Golden yellow

LUSTER: Metallic

MOHS SCALE HARDNESS: 2.5–3.0

SPECIFIC GRAVITY: 19.3 (when pure)

DESCRIPTION AND OCCURRENCE: Gold is a native element mineral that consists predominantly of the element gold. Most gold also contains some silver. When gold contains 20% or more silver, it is referred to as "electrum" rather than gold. Gold can also contain small amounts of other elements, such as copper, iron, bismuth, and platinum group elements.

Gold has a characteristic bright golden yellow color. Gold with a significant amount of silver in it has a paler yellow color. Gold has a metallic luster. Gold is not very reactive, which means that it retains its bright metal appearance rather than tarnishing. Gold is very soft and ductile. Gold is very dense, with a specific gravity that is much higher than most other minerals.

The element gold is found in very low amounts in most rocks. The average abundance of gold in Earth's crust is only 4 parts per billion. That is, for every billion atoms in Earth's crust, only four of them will be gold! However, gold can be concentrated in veins formed by hydrothermal fluids (hot fluids). In these veins, gold is usually found in association with quartz, as well as with pyrite and other sulfide minerals. Gold can weather out of hydrothermal veins into sediments and sedimentary rocks. Because it is much heavier than most other minerals and rocks, gold can become concentrated in sediments, such as stream sediments. When gold concentrates in a sedimentary environment, it forms a placer deposit.

Gold is mined worldwide from both hard rock and placer deposits. Approximately 40% of the gold ever mined has come from South Africa, where it has been mined since the late 1800s from metamorphosed conglomerate rocks in the Witwatersrand Basin. The origin of the gold deposit in the Witwatersrand is not fully understood. The metamorphosed conglomerates probably represent an ancient placer deposit that formed between 2.5 and 3 billion years ago. Ancient microbes may have played a key role in the concentration of large amounts of gold in the Witwatersrand. Other major gold producing countries are the United States, Australia, Russia, Canada, China, and Peru.

In North America, gold mining has primarily taken place during gold rushes. Three of the largest gold rushes were the California Gold Rush, which took place in northern California from 1848 to 1855; the Klondike Gold Rush, which took place in the Yukon region from 1896 to 1898; and the Nome Gold Rush, which took place in Nome, Alaska, from 1899 to 1909. The gold mined during these gold rushes came primarily from placer deposits.

IDENTIFICATION: Gold can be identified by its characteristic golden yellow color, softness, and high specific gravity. Gold can be mistaken for pyrite (fool's gold) or chalcopyrite. However, pyrite and chalcopyrite are much harder than gold and are also much less heavy. In addition, pyrite and chalcopyrite have a dark-colored streak while gold has a golden yellow streak. If you want to look for gold in river or beach sediments, you will want to buy a gold pan. Gently swirling a gold pan filled with sediment and water will concentrate gold and other heavy minerals, such as garnet and magnetite, in the bottom of the pan while letting lighter mineral grains, such as quartz, wash out of the pan.

FUN FACT: One of the largest gold nuggets ever found is the Fricot Nugget, which weighs over thirteen pounds! The Fricot Nugget was discovered in northern California in 1865. Most large gold nuggets have been melted down so that the gold could be sold. However, the Fricot Nugget has been kept in its original form with its beautiful gold crystals still intact. The nugget has toured the world as part of various exhibits, including the Paris World's Fair in 1878. Today, the Fricot Nugget is part of the collection of the California State Mining and Mineral Museum.

TANZANITE

FORMULA: $Ca_2Al_3(SiO_4)(Si_2O_7)O(OH)$

MINERAL CLASS: Silicate

CRYSTAL SYSTEM: Orthorhombic

COLOR: Blue or violet

TRANSPARENCY: Transparent

STREAK: White

LUSTER: Vitreous

MOHS SCALE HARDNESS: 6–7

SPECIFIC GRAVITY: 3.1–3.4

DESCRIPTION AND OCCURRENCE: Tanzanite is a blue- or violet-colored variety of the mineral zoisite, a silicate mineral that is rich in calcium and aluminum. Tanzanite was discovered in 1967 in the country of Tanzania, where it is found in metamorphic rocks. The scientific name for tanzanite is "blue zoisite." However, when the famous jewelry company Tiffany & Co. recognized that the beautiful mineral could be sold as a gemstone, they decided to market the stone under the name "tanzanite." The only known commercial deposits of tanzanite are located in Tanzania, in a small area in the northern part of the country near Mount Kilimanjaro.

Naturally blue or violet tanzanite is very rare, even in the deposits in Tanzania. Scientists are still working to understand exactly what causes the blue or violet color of tanzanite. Most likely, the color is caused by small amounts of the elements vanadium and titanium. When zoisite is heated, the oxidation state of one or both of these elements changes, which leads to a color change in the mineral. A small amount of tanzanite has been produced through natural heating of zoisite. However, nearly all tanzanite gems that are sold in jewelry shops are produced through heat treatment of vanadium and titanium bearing zoisite that was originally another color, such as brown or yellow. Even naturally blue zoisite generally has its color enhanced through heat treatment. Don't worry too much about heat treatment when buying a tanzanite gemstone. The heat treatment is normal and does not impact the quality or value of the stone. However, you may want to ask if a tanzanite gemstone has been treated with a cobalt layer, which is done to improve the blue color of some low quality stones. This cobalt layer can rub off over time, so you will want to avoid it.

To date, no large deposits of tanzanite have been found outside of Tanzania. Geologists are still working to understand how these deposits formed. However, the concentration of vanadium in organic-rich mudstones that were later a source of vanadium-rich metamorphic fluids likely played a key role in the formation of the deposits. Perhaps geologists will find deposits of tanzanite in other countries in the future.

IDENTIFICATION: You are most likely to encounter tanzanite in a jewelry shop or a museum. Look for a gemstone with a striking blue or violet color. The color of the gemstone should look slightly different when viewed from different angles. The color should look more blue or more purple when viewed from one angle compared to another angle.

FUN FACT: When Tiffany & Co. first marketed tanzanite as a gemstone in the late 1960s, the advertising campaign stated that tanzanite could be found in two places, "in Tanzania and at Tiffany's." Today, tanzanite is a popular gemstone and can be found in many jewelry shops.

OPAL

FORMULA: $SiO_2 \cdot nH_2O$

MINERAL CLASS: Mineraloid

CRYSTAL SYSTEM: Amorphous

COLOR: Variable

TRANSPARENCY: Variable

STREAK: White

LUSTER: Vitreous or waxy, sometimes pearly

MOHS SCALE HARDNESS: 5.0–6.5

SPECIFIC GRAVITY: 1.9–2.3

DESCRIPTION AND OCCURRENCE: Opal is an amorphous form of silica that contains water. The water content is variable and is represented by the n in the chemical formula. The water content is most commonly between about 6% and 10%. Because opal is amorphous, it is technically a mineraloid rather than a mineral.

Opal comes in a wide range of colors, including colorless, white, gray, blue, green, red, pink, orange, yellow, brown, and black.

A variety of opal known as "precious opal" is a popular gemstone. Precious opals are iridescent, which means that the colors change as the opal is rotated. This play-of-color is caused by the diffraction of light by silica spheres. The colors produced depend upon the size and spacing of the silica spheres. There are two main types of precious opal: white opal and black opal. White opal has light iridescent colors in a milky white background. Black opal, which is rarer than white opal, has a variety of bright colors (including red, green, blue, and purple) on a black background. A type of opal called "fire opal" has a yellow, red, or orange background color.

A thick piece of opal is required to display the best iridescent effect. Often, opal sold in jewelry shops is something called an opal doublet or opal triplet. Opal doublets are thin slices of opal that have been adhered to a black base that enhances the iridescence. Opal triplets are also adhered to a black base and, in addition, are covered by a piece of plastic or glass, again to enhance the iridescent effect. Opal doublets and opal triplets are less valuable than solid opal gemstones, since only a thin slice of real opal is required for their production.

Opal is generally formed by precipitation from silica-rich groundwater or hydrothermal (hot) fluids. Opal can precipitate in fractures and cavities in any rock type. Nearly all precious opal that is sold for gemstones comes from Australia, where it is mined from sedimentary rocks in several states. In the United States, two locations where precious opal can be found are Spencer, Idaho, and Virgin Valley, Nevada, with both of these localities offering the opportunity for the general public to search for their own precious opals for a fee.

IDENTIFICATION: Precious (or gemstone) opal is easily identified by its iridescence. Common opal is more difficult to identify and can look similar to chalcedony, a silica-rich mineral. However, opal is softer and less dense than chalcedony.

FUN FACT: The world's largest gemstone opal was found in South Australia in 1956. The opal weighs over seven pounds and was named "Olympic Australis" because the Olympic Games were being held in Australia when the gemstone was discovered.

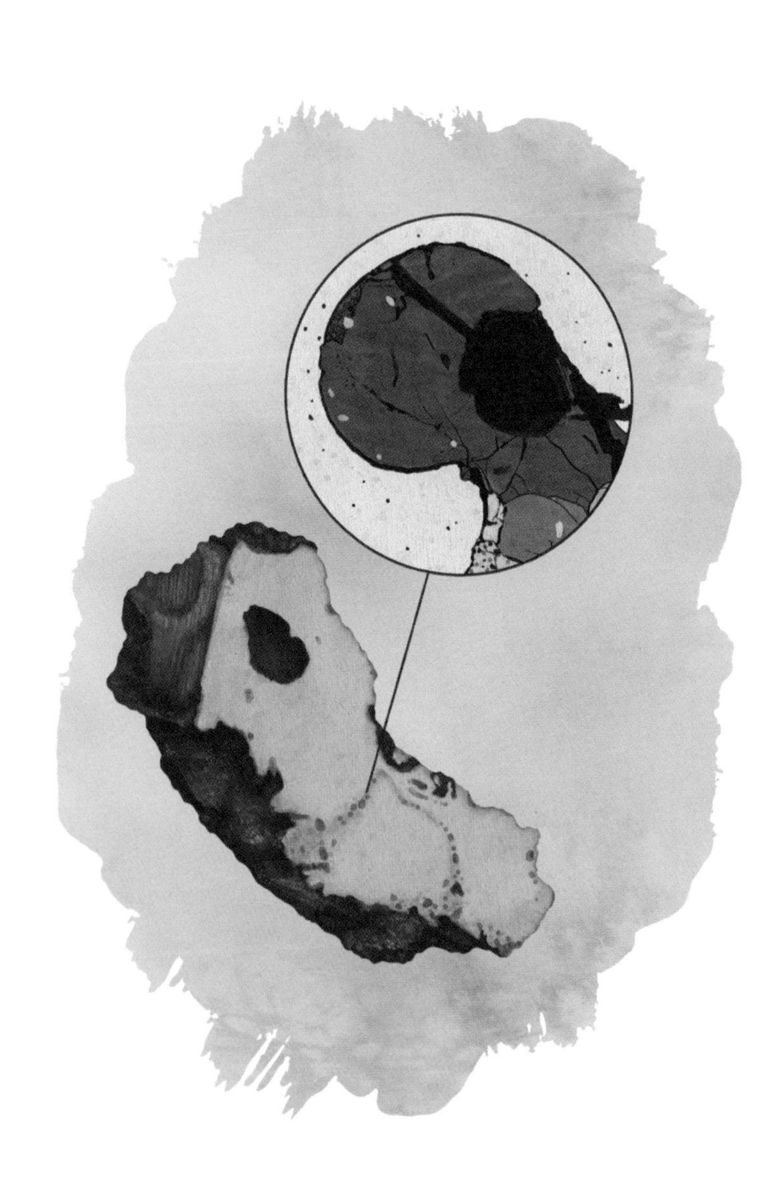

ELKINSTANTONITE

FORMULA: $Fe_4(PO_4)_2O$

MINERAL CLASS: Phosphate

CRYSTAL SYSTEM: Monoclinic

COLOR: Light brown

TRANSPARENCY: Transparent

STREAK: Unknown

LUSTER: Unknown

MOHS SCALE HARDNESS: Unknown

SPECIFIC GRAVITY: 4.2

DESCRIPTION AND OCCURRENCE: Elkinstantonite is one of the rarest minerals known. To date, it has only been found in a single iron meteorite: the El Ali meteorite that landed near the town of El Ali in Somalia. The El Ali rock was known by local people for generations but was only identified as a meteorite by scientists in 2020. The meteorite is very large (it weighs over 15 tons!), but only small pieces of it have been studied by scientists. Recently, a 70 gram piece of the meteorite was sent to the University of Alberta for study. When a team of scientists investigated this sample, they discovered two new minerals. They called the first mineral elaliite, after the name of the meteorite, and they called the second mineral elkinstantonite.

A few small crystals of elkinstantonite were discovered in an inclusion that is surrounded by the iron-rich metal that makes up most of the meteorite. The crystals are only a few tens of microns in size, so scientists have not been able to fully describe all of the mineral properties for elkinstantonite. However, they do know the mineral's chemical formula and that it is a fairly heavy mineral, which makes sense since it contains a significant amount of iron, a heavy element. In addition, they know that the mineral is transparent and has a light brown color.

IDENTIFICATION: Since elkinstantonite has only been discovered (to date) in a single inclusion in a single meteorite, you are highly unlikely to encounter it. Perhaps in the future it will be identified in more pieces of the El Ali meteorite and maybe in other meteorites, too. To identify elkinstantonite, the scientists who discovered it used a powerful microscope called a scanning electron microscope. This type of microscope can take images of very small mineral grains and can also analyze the chemical composition of these grains.

FUN FACT: Elkinstantonite is named after Dr. Lindy Elkins-Tanton, a planetary scientist who has made major contributions to our understanding of how rocky planets form and develop over time. She is a professor at Arizona State University and is also the Principal Investigator for NASA's Psyche mission, which will explore a metallic asteroid that is orbiting the sun between Mars and Jupiter. Psyche may represent the nickel-iron core of a protoplanet (a planetary building block). Scientists believe that Earth has a core that is similar in composition to Psyche, so studying this asteroid will provide valuable clues about our own planet's interior. Since the mineral elkinstantonite was discovered in an iron meteorite, which may also be derived from planetary interiors that broke apart, it is fitting that it was named after Dr. Elkins-Tanton.

RECOMMENDED REFERENCE BOOKS

There are many excellent rock and mineral reference books. A few recommended books are listed below.

EARTH SCIENCE BY EDWARD TARBUCK, FREDERICK LUTGENS, AND DENNIS TASA

This is a textbook designed for introductory university geology courses. The 15th edition of this book was published in 2017. This book is great for someone with an interest in geology but little to no previous knowledge of the subject.

PETROLOGY: IGNEOUS, SEDIMENTARY, AND METAMORPHIC BY HARVEY BLATT, ROBERT TRACY, AND BRENT OWENS

Petrology is the study of the composition, texture, structure, formation, occurrence, and distribution of rocks. This book provides a great introduction to the petrology of all three types of rocks. The 3rd edition of this book was published in 2006.

MANUAL OF MINERAL SCIENCE BY CORNELIS KLEIN AND BARBARA DUTROW

This is a classic mineral reference guide. The first edition of this book was published in 1848 and was authored by James Dwight Dana. This book has been repeatedly updated over the years, with the 23rd edition of the book being published in 2007. For decades, this book has been used as the primary textbook for mineralogy courses at many universities.

THE MINERALS ENCYCLOPEDIA BY RUPERT HOCHLEITNER

This is an excellent comprehensive guide that provides information and photographs for more than 700 different minerals and rocks.

ROADSIDE GEOLOGY SERIES

This is a series of books that has been published by Mountain Press since 1972. The series provides information about the geology of each US state. For example, you can buy books such as *Roadside Geology of New Hampshire* and *Roadside Geology of Southern California.* The books provide a great overview of the regional geology and then give specific information about what rocks you will see when you drive through various parts of the state. Mountain Press also publishes a related series called *Geology Underfoot.* This series provides more detailed geologic information for specific regions, such as for the Yellowstone and Yosemite regions.

AN INTRODUCTION TO THE ROCK-FORMING MINERALS BY WILLIAM ALEXANDER DEER, ROBERT ANDREW HOWIE, AND JACK ZUSSMAN

This is a classic mineral reference guide that has been used by geology students and geologists for many years. Geologists commonly refer to this book as "DHZ" after the last names of the authors. The 3rd edition of the book was published in 2013. The new edition of the book is a little expensive. However, the previous two editions are also very good and can be bought second-hand for very reasonable prices. If you buy one mineral reference book, buy this one.

GLOSSARY

ADAMANTINE: A term used to describe a very reflective mineral luster that is diamond-like.

AMYGDALE: A vesicle that has been filled with a secondary mineral.

APHANITIC: A term used to describe the texture of an igneous rock that has small mineral grains that are not visible with the naked eye.

CLAST: A fragment of a rock or mineral, formed by physical weathering.

CLASTIC: A term used to describe a sedimentary rock that is composed of clasts. Generally, these clasts are visible with the naked eye.

CLEAVAGE PLANE: A plane of weakness along which a mineral tends to break.

CORE: The innermost layer of a planetary body, such as the Earth. The Earth's core consists of an outer core, which is a fluid layer that is about 2,300 km thick, and an inner core, which is a solid ball with a radius of about 1,200 km. The Earth's core consists primarily of iron-nickel metal alloys.

CRUST: The outermost layer of a planetary body, such as the Earth. The Earth's crust ranges from about 5 to 70 km in thickness and consists primarily of silicate rocks.

DULL: A term used to describe a mineral luster that is not very reflective. The term "earthy" is sometimes used instead of dull.

EQUIGRANULAR: A term used to describe the texture of a rock that has mineral grains that are approximately equal in size.

EXTRUSIVE: A term used to describe an igneous rock that forms from the cooling of lava aboveground.

FELSIC: A term used to describe a rock that has a high silica content and has relatively little magnesium and iron compared to a mafic rock.

GLASSY: A term used to describe the texture of an igneous rock composed of amorphous glass.

GRAIN: A mineral crystal or sediment particle in a rock. Terms related to grain size, such as "fine-grained" and "coarse-grained," are often used to describe rocks.

IGNEOUS: A term used for rocks that form from the melting of other rocks.

INTRUSIVE: A term used to describe an igneous rock that forms from the cooling of magma underground.

LAVA: Molten material that flows on Earth's surface.

LUSTER: The way in which light interacts with the surface of a mineral.

MAFIC: A term used to describe a rock that has relatively little silica compared to a felsic rock and has a high content of magnesium and iron. The term "ultramafic" is used to describe a rock with very low silica and very high magnesium and iron contents.

MAGMA: Molten material that is located underground.

MANTLE: A layer located between the crust and core of a planetary body, such as the Earth. The Earth's mantle is approximately 2,900 km thick and is the largest Earth layer. The mantle consists primarily of silicate rocks, mostly peridotite rocks.

METALLIC: A term used to describe a mineral luster that looks like metal.

METAMORPHIC: A term used to refer to rocks that are formed when rocks are subjected to heat and pressure that transforms them into another type of rock, a process known as metamorphism.

MINERAL: A solid, naturally occurring, inorganic material that has a definite chemical composition (for some minerals, the composition varies within a range) and crystal structure.

MINERALOID: A naturally occurring material that is similar to a mineral but which does not have a crystal structure.

NON-CLASTIC: A term used to describe a sedimentary rock that is not composed of clasts, or which is comprised of very small clasts that are too small to be seen with the naked eye.

OBDUCTION: The placement of dense oceanic crust (and often some of the underlying mantle) on top of less dense continental crust. Obduction is somewhat unusual, since dense oceanic crust usually subducts underneath less dense continental crust.

OPHIOLITE: A section of oceanic crust and underlying mantle that has been exposed on land through obduction.

PEARLY: A term used to describe a mineral luster that looks like the shiny surface of a pearl or the inside of a mollusk shell.

PEGMATITIC: A term used to describe the texture of an igneous rock that has very large mineral crystals.

PHANERITIC: A term used to describe the texture of an igneous rock that has mineral grains that are large enough to see with the naked eye.

PHENOCRYST: A relatively large crystal located in a finer-grained matrix in an igneous rock.

PLACER: A mineral deposit, for example of gold, that forms in a sedimentary environment.

POLYMORPH: A mineral with the same chemical formula but a different crystal structure as another mineral.

PORPHYRITIC: A term used to describe the texture of an igneous rock that has larger crystals (phenocrysts) in a finer-grained matrix.

PORPHYROBLAST: A large mineral crystal in a finer-grained matrix in a metamorphic rock.

ROCK: A naturally occurring solid mass that is comprised of one or more minerals or mineraloids.

SEDIMENTARY: A term used for rocks that are composed of fragments of rocks, minerals, and/or organic materials that become cemented together (non-chemical sedimentary rocks) or from minerals that precipitate from solution (chemical sedimentary rocks).

SILKY: A term used to describe a mineral luster that looks like silk. Generally, fibrous minerals are described as having a silky luster.

SUBDUCTION: The movement of one tectonic plate underneath another, for example dense oceanic crust underneath less dense continental crust.

TECTONIC PLATE: A piece of the Earth's crust and upper mantle. The Earth is divided up into a number of tectonic plates, which move slowly over time and drive a number of geologic processes, such as mountain building and the generation of new oceanic crust on the seafloor.

VESICLE: A small hole or cavity left in a volcanic rock by escaping volcanic gases.

VITREOUS: A term used to describe a mineral luster that looks like glass.

WAXY: A term used to describe a mineral luster that looks similar to wax.

XENOLITH: A foreign rock inclusion, usually in an igneous rock. In igneous rocks, xenoliths are generally picked up by magmas as they ascend to the surface and may represent portions of Earth's mantle or crust. The word "xenocryst" can be used in a similar way to describe a foreign crystal.

ACKNOWLEDGMENTS

First and foremost, I must thank my friend Dr. Jay Barr. Jay kindly photographed many rocks and minerals in his collection to provide reference photographs for the illustrations. In addition, Jay read through the rock and mineral profiles, checking them for accuracy and offering many useful suggestions to improve the writing. However, I must take responsibility for any scientific errors that may have slipped into the book.

The staff at Queensland Museum are thanked for providing reference photographs of rocks and minerals. In addition, I must thank Dr. Thomas Jones for arranging access to The University of Queensland Geology Museum so that I could photograph the specimens on display there. I must also thank my friend Dr. Lisa Falk for providing additional reference photographs. Dr. Bruce Cairncross provided the reference photograph for tanzanite and the copyright for that image remains with him. Dr. Hugo Olierook and Dr. Janne Liebmann are thanked for providing the reference photographs for zircon. Dr. Chris Herd and Dr. Chi Ma are thanked for providing the reference meteorite hand sample and microscope images for elkinstantonite. I must also thank Dr. Chris Herd and Dr. Lindy Elkins-Tanton for reviewing the elkinstantonite mineral profile.

Many people have contributed to my knowledge of geology, which I first started studying more than twenty years ago. However, I would like to mention two people who have been particularly inspiring during my study of geology. The first person is Alice Hallaran, who taught the first geology class that I ever took at Westover High School in Connecticut. Alice encouraged my interest in rocks and minerals. Soon after, I started studying geology at university. The second person is Dr. Susan Humphris, who was my Ph.D. supervisor at Woods Hole Oceanographic Institution. Susan is an incredible scientist and provided me with unwavering technical and emotional support during my doctoral studies. I feel incredibly fortunate to have had the opportunity to work with Susan.

I would like to thank Lindy Pokorny, Jordan Stockman, and the rest of the team at Cider Mill Press for inviting me to write this book and for all of their hard work putting the book together. I would also like to thank Vlad Stankovic for creating the amazing rock and mineral illustrations.

Last but certainly not least, I must thank my wonderful family: Jackie, Caspian, Mom, Dad, Aynsley, and Morgan. My husband Jackie is also a geologist, and we have had many geological adventures over the years. I hope that we can continue to look for rocks and minerals around the world for many years to come.

ABOUT THE AUTHOR

Dr. Evelyn Mervine is a geologist and climate change expert. She has a B.A. degree in Earth Sciences from Dartmouth College, an M.Sc. degree in Carbon Management from The University of Edinburgh, and a Ph.D. degree in Marine Geology that was jointly awarded by The Massachusetts Institute of Technology and Woods Hole Oceanographic Institution. She has worked as a geologist all over the world, including in North America, the Middle East, Africa, and Australia. She writes a geology blog called Georneys. Find her on Instagram @evelyn_georneys.

ABOUT CIDER MILL PRESS BOOK PUBLISHERS

Good ideas ripen with time. From seed to harvest, Cider Mill Press brings fine reading, information, and entertainment together between the covers of its creatively crafted books. Our Cider Mill bears fruit twice a year, publishing a new crop of titles each spring and fall.

"Where Good Books Are Ready for Press"

501 Nelson Place
Nashville, Tennessee 37214
cidermillpress.com